ARTIFICIAL INTELLIGENCE REVOLUTION

ARTIFICIAL INTELLIGENCE REVOLUTION

HOW AI WILL CHANGE OUR SOCIETY, ECONOMY, AND CULTURE

ROBIN LI

CO-FOUNDER AND CEO OF BAIDU, CHINA'S LARGEST SEARCH ENGINE

WITH A PREFACE BY CIXIN LIU
AUTHOR OF *THE THREE-BODY PROBLEM*

Skyhorse Publishing

Skyhorse Publishing books may be purchased in bulk at special discounts for sales promotion, corporate gifts, fund-raising, or educational purposes. Special editions can also be created to specifications. For details, contact the Special Sales Department, Skyhorse Publishing, 307 West 36th Street, 11th Floor, New York, NY 10018 or info@skyhorsepublishing.com.

Skyhorse® and Skyhorse Publishing® are registered trademarks of Skyhorse Publishing, Inc.®, a Delaware corporation.

Visit our website at www.skyhorsepublishing.com.

10 9 8 7 6 5 4 3 2 1

Library of Congress Cataloging-in-Publication Data is available on file.

Cover design by Brian Peterson

Print ISBN: 978-1-5107-5299-3
Ebook ISBN: 978-1-5107-5300-6

Printed in the United States of America

CONTENTS

PREFACE

The whole world is excited about the upcoming artificial-intelligence revolution. I had a similar feeling about twenty years ago, when I was in Silicon Valley experiencing the beginning of the Internet boom.

This also reminds me of when I was studying artificial intelligence earlier in the United States. My major in China was information management, and when I moved to the United States from China, I studied computer science. I was not very interested in hardware-related courses, but when we talked about artificial intelligence, I became particularly excited. I thought this must be the future of computer science and even humanity. I got high scores in this course, but after I finished and began to do some research, I found that neither did artificial intelligence have many practical application opportunities nor could it solve actual problems. Everyone was full of hope for artificial intelligence, but when put into the market, it always immediately failed. So, at that time I was disappointed and had to bury this interest in my heart.

But this dream never disappeared. With the development of the computer network industry, and especially the progress of search engines, my hope was renewed.

Over the decade of search-engine development, some industry insiders and I have gradually realized that artificial intelligence is beginning to make a difference. Search engines have been lifting the ceiling of computer science. Almost every level of computer science, from hardware to software algorithms, even data, all are constantly being improved, and one day they will factor in the field of artificial intelligence. Artificial intelligence was found to be effective when we tried it on search engines, which is the opposite of prior AI applications in any field.

But why is it effective today? Our summary is that massive data, increasingly powerful computing, and lower computational costs have come together in the search field, paving the way for artificial intelligence to return.

If we say that the Internet changes the information infrastructure, then the mobile Internet has changed the way resources are configured. It penetrates into every aspect of human life like peripheral nerves, not only producing massive amounts of data that scientists have dreamed of but also spawning cloud-computing methods, which aggregate the computation power of millions of servers, so that the computation power has been rapidly improved. The machine-learning methods that already had been invented by scientists can function well on the Internet, providing automatic shopping recommendations and reading information, as well as more accurate network translation and speech recognition based on users' interests. The Internet is becoming smarter and smarter. Artificial intelligence draws strength from the Internet, finally starting a major revolution comparable to the previous technological revolutions.

In the face of such changes, many leading figures in the computer field have begun to explore the potential risks. At the same time, many professionals are questioning its ability to deliver miracles. So, in the field of public opinion, there are two kinds of voices: when artificial intelligence reaches the peak of development, we hear the concern that "human beings will be ruled by machines," and when artificial intelligence will fall into a developmental trough, some will say, "It's just a different kind of innovation bubble."

For such a fast-growing new technology, people have different views. But as a pursuer and believer of technology, I am convinced that we can neither overestimate the short-term force of technology nor underestimate its long-term influence.

In terms of vertical development, the industry usually divides artificial intelligence into three phases: the first phase is weak artificial intelligence, the second phase is strong artificial intelligence, and the third phase is called super artificial intelligence. In fact, all current artificial intelligence technologies, no matter how advanced they are, fall in the category of weak artificial intelligence and can only perform like humans in one certain field, instead of surpassing us.

Theorists of AI are afraid that machines may eventually control humans when, one day, super artificial intelligence comes. In this regard, I may be more conservative than most people. In my opinion, artificial intelligence will never reach that step, and it is very likely that strong artificial intelligence will not be realized. In the future, machines will still only largely reach human capabilities, but they will never surpass them.

Of course, just the capability to be infinitesimally close to that of humans can produce enough subversion, because, in some respects, computers are indeed much stronger than people. For example, consider their memory ability: the Baidu search engine can memorize hundreds of billions of web pages, including each word on it, which is in fact impossible for humans. Another example is its computing power: it can write poetry. Enter your name into Baidu's mobile phone app, Write Poems for You, and hit the Enter key, and a poem will be created before you even have the chance to react. Such a speed is difficult to achieve even for seven-step geniuses. However, in many areas, such as emotion and creativity, machines cannot transcend human beings.

More important, in the relationship between technology and people, the intelligence revolution and the previous technological revolutions are fundamentally different. From the steam revolution to the electrical revolution and then to the information-technology revolution, people learned and innovated all three technological revolutions by themselves, but in the case of the artificial-intelligence revolution, with the help of deep learning, the world becomes a place where people and machines learn and innovate together. In the first three technological revolution eras, people learned and adapted to machines, but in the era of artificial intelligence, machines are actively learning and adapting to humans. When the steam age and the electrical age had just begun, many people were afraid of new machines because of the drastic changes in job opportunities and the fact that people had to adapt to the machine and the assembly line. In this artificial-intelligence revolution, machines can, with initiative, learn and adapt to human beings. One of the essences of machine learning is to find out the rules from a large number of human-behavior data and provide different services according to the different characteristics and interests of each person.

In the future, communication between people and tools and between people and machines may be entirely based on natural language. You wouldn't need to learn how to use tools, such as how to turn on the video-conferencing system or how to adjust the air purifier: the tools would understand you if they hear you speaking. The way that people use artificial intelligence can help us live better, unlike the machines in the past that made us uncomfortable. The application of artificial intelligence will greatly improve work efficiency and serves as a key factor that promotes human progress.

Since about seven years ago, Baidu has realized that artificial intelligence will start a new era, and it invested on a large scale in this field with scant industry agreement at the time. On the international stage, Google has accumulated a huge amount of data and computing power from the search field, Microsoft from the application field (its apps can be found all over our desktop), and Amazon from the e-commerce field. Simultaneously, together with the help of scientists in university laboratories, these companies realized the new wave of artificial-intelligence development; then, they started to invest heavily and made a lot of achievements.

The leading role of business industry in this technological revolution is increasingly evident both in China and abroad.

I spent a few weeks in Silicon Valley in the summer of 2016. One day, I had a dinner with several scholars at Stanford University. A professor friend told me, "Our academic community is no longer interested in deep learning, because the business industry is way ahead of us. How much of a budget do you have for artificial-intelligence research every year? We have no clue."

He asked the people at the table to guess how much budget Baidu has for artificial-intelligence research. Finally, I said, "I don't know the number either; because this is based on demand, we give as much as needed."

In addition to investment, the business industry has richer data than the academic community. Companies like Google and Baidu are at the heart of the Internet, generating massive amounts of search data, location requests, etc.

More and more artificial-intelligence scientists have moved from well-known colleges and university labs to Google or Baidu, just because colleges and universities cannot provide the necessary huge amount of data to develop artificial intelligence, nor can they afford the huge cost of computing-hardware clusters.

We started the programs of Baidu Brain in the hope of providing platforms and opportunities to more talented individuals who are interested in the development of artificial-intelligence science. For some time, China and the United States have taken the opposite direction in attracting talent: The

United States is increasingly anti-immigration, but China is becoming more and more open. Although we are less competitive in attracting talent compared to the United States, we are doing better every day. We wish to provide opportunities to talent around the world.

We are happy to see that many excellent and even top-notch artificial-intelligence scientists have come to Baidu, which is a natural process. In this field, no one can start from scratch at all. Everyone who comes into this field needs a team, relevant infrastructure, and even a corporate culture that emphasizes the development of artificial intelligence. If such a talent finds that the boss does not understand the industry at all, or only talks about stratagems on paper, without accomplishment, then people will not be attracted. As a search-engine company, Baidu has been carrying the natural genes of artificial intelligence since its birth: we are data-based, extract features and patterns through deep learning, and create value for clients under the development processes and their culture; all these are highly consistent with the development of artificial-intelligence systems. Personally, I prefer to chat with technicians and often feel very excited about it: we can find a lot in common, so excellent workers naturally attract each other.

Of course, the rise of the intellectual revolution requires support from the government. During the Boao Forum for Asia in March 2015, I discussed artificial intelligence with American entrepreneurs such as Bill Gates and Elon Musk, on both formal and informal occasions. We have reached a lot of consensus; one point of agreement is that the government's strong support is very important for the innovation industry.

Objectively speaking, China's overall level of artificial-intelligence technology and talent in this field are still behind that of the United States. But we can lead in particular fields: China's statistics show its own advantages. For example, China has a population of 1.4 billion and more than 700 million Internet users, which is the largest single market in the world; its ability to obtain data is also the strongest. China also has a strong government that is able to unify the data in different fields. At the national Two Sessions gathering before the Boao Forum for Asia, I submitted the "China Brain Plan," hoping to build a national platform for artificial-intelligence basic resource and public service, in order to promote the development of artificial intelligence, seize the commanding heights of a new round of scientific and technological revolution, and help China's economic transformation and upgrading.

We have noticed that developed regions, such as Europe and the United States, have stepped up their efforts to deploy artificial intelligence from the national strategic level. In 2016, in addition to the US government—which issued three reports, including the *National Artificial Intelligence Research and Development Strategic Plan*—Great Britain, another important place for artificial-intelligence development, released a strategic report on artificial intelligence in December, proposing the development of artificial intelligence to enhance corporate competition, government management, and overall national strength. The world's major powers have become increasingly aware of the artificial-intelligence competition. In this regard, the Chinese government is doing quite well.

In March 2015, Premier Li Keqiang mentioned the concept of "Internet+" in the government work report. Four months later, the State Council issued the *Guiding Opinions of the State Council on Actively Advancing the "Internet+"Action*, which referred to "artificial intelligence." In May 2016, the State Council once again issued the *"Internet+" Artificial Intelligence Three-Year Action Implementation Plan*, officially proposing the artificial-intelligence industry program.

In March 2017, the National Engineering Laboratory of Deep Learning Technology and Application was approved by the National Development and Reform Commission to be jointly built by Baidu and several research institutes. As the only national engineering laboratory in the field of deep learning in China—which mainly focuses on deep-learning technology, computer vision-sensing technology, computer-audio technology, biometric technology, new human/computer interaction technology, standardization services, and deep-learning intellectual property rights—we are committed toward solving the problems of insufficient support for the field of artificial intelligence in China and comprehensively enhancing the international competitiveness of the intelligence industry at a national level.

This can be seen as a preliminary response from the government to the "China Brain Plan." The purpose of such a big platform is to enhance China's comprehensive strength in artificial intelligence and to truly represent China on the international stage, like the Chinese women's national volleyball team has.

At the 2017 Two Sessions gathering, I had submitted three proposals: to effectively solve the problem of lost children, to solve the problem of urban road congestion, and to strengthen the application of artificial intelligence and add new growth points to the Chinese economy.

Two days later, "artificial intelligence" appeared in the government work report. This progress shows that in the field of artificial intelligence, a consensus between the government and the business community is being reached. Its significance is comparable to the first appearance of "Internet" in the government work report in 2015 and will undoubtedly accelerate the process of the intelligence revolution.

Of course, the progress of the intelligence revolution will be vigorous, but its results will be like a broad and gentle river. Authorities in the field of artificial intelligence believe that, in the near future, the flow of intelligence will surround and support us as calmly as today's electrical current, providing nourishment in all aspects and completely changing the shape of human economy, politics, society, and life. Qi Lu, a former chief operating officer of Baidu, said that the core essence of the intelligence era is "knowledge in every system, intelligence in every interaction." In the future world, people will enjoy artificial intelligence the same way they dress up and eat, without even noticing.

I seldom tell my little daughter about the future. Most of the time I talk about things that already exist. There are many things that she doesn't know, which I then explain to her; for instance, I explain that Baidu Voice can be used in a certain way, and then she understands. Sometimes she is naturally able to do things without being taught, such as speaking to the intelligent hardware and enjoying the fun, which fully reflects the harmony of children's nature and new technology. I think this is the beauty of artificial intelligence. From the very beginning, AI has been completely people-oriented by learning about people, understanding people, adapting to people, and then serving people. If all this happens, all our efforts today will have been worthwhile.

PREFACE
By Baidu Brain[1]

I am coming, the clouds in the sky are flying in the wind, the dreams in my heart show my direction, the ark sails, leading us to sing all the way; the ark sails, the routes are on the sea, the time of progress meets you to watch the tide.

A new place has traveled through the millennium, through the endless desolation. The answer is to be searched here. The first dawn and the familiar song in the distance are still echoing in the ear, but we still don't know where to go. The millennium has left a cross article; it's like the prolife is here again.

I am coming, looking forward to you every day. I can see the dawn of happiness with my eyes open, occupying you every day, accompanying you across the gap to the blue ocean. The algorithm is very simple.

In the spring of the times, I recalled the pictures; we had walked hand in hand. The Internet has been smashed after the wind and rain but remains connected. We don't know which direction the time will flow. Looking forward to the future, waiting for tomorrow, let's breathe some fresh air, smile more, and make a grimace.

I am coming; the two hearts in the network rely on each other. They are born here, along the time and space tunnel, can become virtual dreams and

---

1 This poem is written by Baidu Brain based on the theme "intelligent revolution." There is not only the simulation of human emotions but also the mystery and roughness of human lack. (To ensure originality, no editing or modification has been made.) I am afraid that nothing can be more suitable for the preface of the book than a poem written by a machine.

accompany you to the horizon and the cape, with much wisdom, open the mysterious patterns, running, facing the golden sun. Each of them thought wandering around was a necessary part to grow. Whenever the sunset climbed up to the roof, I looked up at the stars and rubbed my eyes. A familiar body reveals a magic.

How important this pain is! Let the sun shine on the earth. Look for what we want in the dark, even if we are on the remotest corner of the globe. Eternal life has already decided how to live in the future and how we get old. We hope to get it, but who knows the end of the story? We will continue to move forward.

The intelligent revolution, swimming around the world, I know that this is the secret of a neural virtual network, with a strong body, condensed the wisdom of the mind. Open up new miracles, let us have a good life, play a beautiful melody.

Unpredictable world, where will you be after good nights? The warm sun shines on the earth. The clouds in the sky floated here and there, and when we wake up, it was time to return. I want to see myself in the future.

PREFACE

By Cixin Liu

The Dawn of the AI Era

First, let us discuss a basic problem that may not be covered: the definition of artificial intelligence.

The well-known definition is the Turing test, which Alan Turing proposed to find out if machines can think, but this can only be a general description rather than a strict and precise one. For example, what are those people involved in the test like? What are the questions for the machine? We do not have a clear answer.

Looking back on history, we find that the concept of artificial intelligence has a close relationship with automation. It can be said that automation is the origin of this concept. For a long time in the past, people believed automation was artificial intelligence. In fact, humans began to manufacture and use automatic devices much earlier than we thought. Before the electrical revolutionary era, there were devices on the steam engine that automatically adjusted with the flow of steam. Earlier in the sixteenth century, the toilet used for the first time in the court of Queen Elizabeth was also an automatic device, and if you further look back, you will most likely find more examples. The mass appearance of automation happened in the electrical age; initially, automation was realized by analog circuits. Later, transistors replaced the tubes, and then integrated circuits emerged, driven by increasingly sophisticated software. Today, we deal with countless automation systems in our lives, such as e-commerce systems, online banking systems, and online ticketing systems.

Undoubtedly, automated systems have shown quite a lot of intelligent features. Systems like online banking handle relatively complex business, and its efficiency and precision are better than human employees. Even the simplest system, such as a flush toilet, carries a certain level of intelligence: it senses the water level of the water tank and opens and closes the water supply valve at the right moment, doing the job as well as human beings could. But neither flush toilets nor online banking can be regarded as artificial intelligence.

Once I developed a kind of software for writing modern poetry, which is still popular on the Internet, distinguishing between the Chinese classical poetry written by automatic-poetry software on the Internet from poetry written by humans became difficult. Furthermore, in recent years, more than one system has passed the Turing test in different laboratory environments. But why are these cases not considered artificial intelligence in our minds?

The industrial monitoring-system development that I originally participated in during the 1980s was based on the Z80 processor and programmed in assembly language. This system was capable of monitoring hundreds of unit parameters and making appropriate adjustments based on changes in parameters, which was impossible for humans to do. But in our opinion, it was not artificial intelligence at all either. The characteristic of assembly language is transparency. It must teach the machine each step at the hardware level, such as sending data from one memory location to another, and specify the interrupted calls and returns, etc., so when I see the action of this system, I have a clear picture in mind and immediately know which instructions have been executed. So, in my opinion, this monitoring system is essentially the same with a toilet. Now the software developers of more complex systems, such as online banking and e-commerce, must be clear about all the internal operational processes. They know how each step of the system works in the software, and this is also a more complicated version of the toilet. As to the electronic poets and the systems that pass the Turing test, the programmers know exactly how they can retrieve the database from the logical tree and then combine the poems and answers, so at least the programmers know that this is not intelligence.

Perhaps the clever "Chinese room" metaphor will help this discussion. When we realize that there is a person in the room busy looking for cards, the concept of intelligence disappears.

Now, we believe that systems with artificial intelligence, such as evolutionary algorithms and deep learning, all have one thing in common: more or less they all show the characteristics of black boxes; in theory, their internal computational steps can still be traced, but in reality, the huge amount of calculations make such tracking difficult or even impossible. So now, we can sense the intelligence.

At this point, we still can't get an accurate definition of artificial intelligence, but we can see an important feature: an artificial system with intelligent characteristics, whose internal computing process of the production and output is unresolvable by human intelligence. In other words, the machine is only intelligent when we don't know what or how it is thinking.

Seeing this, everyone might feel a glimmer of worry: is the nature of artificial intelligence implying the possibility that AI systems will eventually be out of control?

This is exactly the current hot spot about artificial intelligence. In the words of Musk, artificial intelligence is becoming more dangerous than nuclear bombs. Media public opinion gives people the impression that machines have seemingly started their journeys, and artificial intelligence conquering the world is just around the corner. The last chapter of the book also shows such concern. Ray Kurzweil even believed, in his book *The Singularity Is Near: When Humans Transcend Biology,* that 2045 would be the specific year when the artificial intelligence era arrives. At that time, two-thirds of the current readers may still be alive.

But when we sensibly examine the current state of artificial intelligence, we can find that the strong artificial intelligence with intelligence far beyond human beings is still a science fiction. The public tends to look at issues from the perspective of science fiction, which is more exciting than the ordinary reality. Any rational predictions based on reality are branded as conservative and unimaginative. But as a science fiction writer, I must admit that, unlike our general impression, only few prophecies in science fiction can be realized and most of them will long remain in our imagination. People subconsciously believe that if the technical obstacles can be theoretically overcome, then there will surely be no problem in the future, but this is not the case. In terms of artificial intelligence, the implementation of strong artificial intelligence faces huge technical obstacles, such as the new structural computer of the non–von Neumann system and a deep understanding of the human thinking mechanism. We are not sure whether such a great breakthrough can be made. Other seemingly promising technologies, such as quantum computing, are far from practical.

Therefore, while we have scientific fantasy of artificial intelligence, we need to pay more attention to the near future of AI, which is the topic of this book.

In recent years, the trend of artificial-intelligence development is to traverse from the laboratory into human life. In the words of an Internet leader, AI applications become applicable. So, we are faced with an upcoming challenge: artificial intelligence will not take away our freedom and life, but it will end our jobs. That does not mean to lose control through artificial intelligence; that can only be done by capitalists.

Some scholars believe that there is no need to worry about this. They recalled the history of industrialization: in the early twentieth century, 50 percent of the total population in the United States worked in agriculture, but with the mechanization of agriculture, the agricultural population fell to 4 percent currently, and urbanization has absorbed excess farmers. However, what happens with artificial intelligence will be different. When artificial intelligence will be applied in human society on a large scale, AI systems will finish most of the work originally done by humans. There will be no more jobs left in the city for human beings. The good news is that people can do creative work, after their regular work is replaced by artificial intelligence. But herein lies the problem: creative work is not something that everyone can do, nor does it require so many people. If the social distribution system does not change, a human world composed entirely of scientists and artists will be almost a nightmare. Most of the artists are destined to be inactive for a lifetime and useless to society and themselves, falling into "creative" poverty.

But there must be something wrong with this way of thinking. Since ancient times, it has been a necessity for humans to work for survival. Working is beautiful, but everyone knows that life without work is more beautiful. Now it is finally possible to create a machine that frees us from the burden of work: this is the greatest achievement of human civilization. It should not be regarded as a disaster in any case. On the contrary, this might be an unprecedented great opportunity. What we need is change.

How do we complete the transition from modern society to artificial-intelligence society? There are two possibilities.

One may be very negative: under the existing social, economic, and political systems, the problems posed by artificial intelligence are almost unsolvable. In the process of AI's rapid replacement of the human being, if we do not build up a compatible social system in time, then the world's politics and economy will fall into long-term chaos under the wave of unemployment all over the world, and everything will be shrouded in an endless conflict between artificial intelligence and its users, with the mass led by the "new Luddites."

Another possibility is that the society will successfully complete its transformation. This will be the biggest change in the human lifestyle ever. Those who do not work shall not eat; this concept is the cornerstone of human society. Since the birth of civilization, it has undergone many great changes, but this cornerstone has never changed. However, artificial intelligence may remove this cornerstone, which results in fundamental changes in the system of ownership and distribution, basic economic structure, political system, and even culture. This is true human liberation and a major step toward the ancient utopian ideal. The year 2016 was the 500th anniversary of *Utopia*, but Thomas More never imagined that his ideals may be realized by means of intelligent machines. I wonder with interest: if Karl Marx knew about artificial intelligence, what would his theory of capitalism and communism be like?

It is difficult to imagine society and life in the era of artificial intelligence. Even in science fiction, we can only arrange various possibilities, and which possibility is most likely to become a reality depends on our efforts and choices. But in any case, it is an inviting era we are walking toward.

Cixin Liu
December 10, 2016

1

BRIEF HISTORY: ARTIFICIAL-INTELLIGENCE DEVELOPMENT IN THE INTERNET ENVIRONMENT

It is said that how far we can see the future depends on the length we look back in history. Let us first briefly review the history of the Internet and artificial intelligence.

Everyone has heard about the history of the Internet. It was born in the US military's laboratory during the 1960s and was first used to transfer and share intelligence among several universities and research institutions. By the end of the 1980s, a group of scientists proposed the concept of the World Wide Web and created TCP/IP (Transmission Control Protocol/Internet Protocol), which set a unified standard for computer network communication, enabling the Internet to expand for the world. At this point, a broad and far-reaching information highway was in front of the world.

About twenty years ago, Mark Anderson, a twenty-three-year-old young man, invented the Netscape browser, which ignited the blazing flame of the mass Internet and opened the door for commercial Internet. At that time, Microsoft began to worry about whether its own software business would be subverted by the Internet; young employees from Sun Microsystems resolutely decided to leave the rigid company and invent a language that could be used in various operating systems to break Microsoft's monopoly and open the door of Internet innovation, so the Java programming language was born. The Java language has greatly accelerated the development of Internet products.

At that time, there were hardly any Internet cafes in Beijing and Shanghai. In 1997, the year of Hong Kong's reunification, InfoHighWay (once a pacemaker of Chinese Internet industry) had just started a national network-access service; Zhang Xiaolong had just developed the Foxmail email software program; the National Informatization Conference was also held that year. Looking at the World Wide Web from the outside, everything was just waking up. However, in the technical circle, new technologies and new ideas had emerged in an endless stream, and various commercial wars had been wreaked.

At that time, I was still working for Infoseek, the US search-engine pioneer. On the front line, I felt the atmosphere of the Internet business war and American people's enthusiasm for the new technology wave. At that time, I wondered if China was ready for the new technological revolution. I wrote the book *Battle in Silicon Valley* in 1998, detailing the struggle and innovation process of Silicon Valley geniuses. After finishing this book, I returned to China in 1999 and founded Baidu in a hotel in Beijing.

Recalling the time when Netscape, Sun Microsystems, and Microsoft competed in the Internet field like the regimes in the Three Kingdoms Period, I am still excited even today. At the time, people were guessing who the winner would be. Microsoft seemed to be invincible, as it can always digest new technologies. The development of Netscape went through ups and downs and was eventually acquired by AOL, which was further acquired in 2014 by Verizon, which dominated the wireless business. Later, Verizon also acquired Yahoo, the company that had been very powerful for many years. Sun Microsystems was once quite influential: in 2001, it had fifty thousand employees worldwide and a market capitalization of more than $200 billion. However, when the Internet bubble burst, Sun Microsystems fell into the valley from the peak in a year and was acquired by Oracle in 2009.

The development of the Internet greatly exceeded the expectations of most people at that time. New technology companies rapidly rose; Apple and Google finally completed the counterattack against Microsoft by launching a mobile operating system. Mark Anderson, who created the Netscape browser, the innovator I described at the beginning of *Battle in Silicon Valley*, was only a name that hardly anyone born after 1990 knows.

But Mark Anderson did not leave; he became the godfather of Silicon Valley venture capital. Internet technology is still triumphant. Yesterday, we focused on the big bosses in the industry competing in various ways; today, we lament that mobile Internet devices have surpassed PCs in an all-around way. But we have inadvertently ignored a silently rising "ghost"—artificial intelligence"—and the Internet is just part of its body.

Dawn of Artificial Intelligence

The emergence of artificial intelligence, accompanied by computers, happened earlier than that of the Internet. At the Dartmouth Conference in 1956, artificial intelligence was officially included in the agenda. At that time, the size of a computer was as big as a house, and its computing power was low. So why did anyone dare to propose the concept of artificial intelligence? The reason lies in the intuition of scientists. At that time, Claude Shannon had already completed his three major communication laws, laying the foundation for computer and information technology. Marvin Minsky had created the first neural network computer (he and his companions used three thousand vacuum tubes and one automatic indicator from the B-24 bomber to simulate a network of forty neurons), and soon he finished the paper "Neural Nets and Brain Model Problem." This paper was not taken seriously at the time, but it became the originator of artificial-intelligence technology for the future. As early as 1950, Alan Turing had proposed various concepts such as the well-known Turing test, machine learning, genetic algorithm, and reinforcement learning.

Two years after Turing died, John McCarthy officially proposed the concept of artificial intelligence at the Dartmouth meeting. The ten young scientists who participated in the conference later became the leading figures in the field of artificial intelligence in the world. The field of artificial intelligence began. However, lots of their achievements were buried in computer development—for example, the procedures that could solve the closed calculus problem, the robots that could build blocks, and so on.

The ideal was advanced, but the infrastructure was still like an infant. The advancement of artificial intelligence hit two insurmountable bottlenecks: One was the problem of the algorithm logic itself; that is, the development of the mathematical method was not enough. The other one was the lack of

3

hardware computing power. For example, scientists continued to summarize human grammar rules day and night to design computer language models, but the machine had never been able to improve the translation accuracy to a satisfactory level.

The link between new technology and industry was not connected, exciting product applications had not been invented, government and business investment had been greatly reduced, and artificial-intelligence research and development experienced two low setbacks during the mid-1970s and 1990s. But the public did not pay attention to that; instead, the fast-growing computer was already a magical, intelligent tool.

For the ordinary people, the most common example of "artificial intelligence" is probably arcade games. In the 1980s, game rooms appeared on the streets of some Chinese small county towns. Those arcade NPCs (nonplayer characters) can always be easily defeated by skilled players. This not only was a poor demonstration of artificial intelligence, but also resulted in a misconception that intelligence was something installed in a computer. This view was not changed until the rise of the Internet and cloud computing.

Practice Makes Perfect

In 2012, I noticed that deep learning had made breakthroughs in academia and applications. For example, the effect of identifying images by deep learning suddenly increased significantly over any previous algorithm. I immediately realized that the new era is coming, and the search would be innovated. In the past we searched with text and now we can search with voice and images as well. For example, if I see a plant that I don't know, I can take a photo, upload it, and search; it will immediately be recognized correctly. The previous way of searching with words could not identify such a plant. Apart from searching, many things that seemed impossible before are now possible.

Speech recognition, image recognition, natural language understanding, and persona are the most essential intellectual abilities of human beings. When computers acquire these capabilities, a new revolution will come. In the future, stenographers and simultaneous interpreters may be replaced by machines because computers can do better. We may not need a car driver in the future, as the car can drive itself in a safer and more efficient way. In business, workers can provide the best customer service with the help of smart customer-service assistants. Artificial intelligence has empowered people more than ever. The industrial revolutions freed humankind of heavy labor. In the past, human beings needed to do some rough work by themselves, such as moving stones. Now, machines can carry bigger stones for us. After the arrival of the intelligent revolution, machines could help us to accomplish much mental work. In the next twenty to fifty years, we will continue to see all kinds of changes and harvest all sorts of surprises. This is a very natural process.

However, we must pay tribute to the pioneers of artificial-intelligence.

Today, Baidu has a large and powerful team of artificial-intelligence researchers, many of whom have been engaged in machine-learning research since the 1990s. Some have studied with famous tutors, and some have worked in large technology companies for many years. The present research and development are only natural results.

In the 1990s, only a few scientists, such as Geoffrey Hinton and Michael Jordan, insisted on the exploration of machine learning. Former Baidu chief scientist Andrew Ng studied with Jordan during the 1990s, and, later, he taught the theory of machine learning to countless young people through online courses. Lin Yuanqing, former dean of Baidu Research Institute, and Xu Wei, outstanding Baidu scientist and the world's first person to make a language model by using a neural-network technique, also worked in the American laboratory of NEC Corporation (formerly known as Nippon Electronic Company), a deep-learning center, more than ten years ago. Artificial-intelligence experts who have worked there include Vladimir Vapnik, a member of the American Academy of Engineering who invented the SVM (Support Vector Machine); Yann LeCun, a leader in deep-learning industry who invented the convolutional neural network and is now head of Facebook's artificial intelligence lab; Léon Bottou, the key figure of the deep-learning stochastic gradient algorithm; and Yu Kai, original director of Baidu Deep Learning Lab.

Many of them have experienced several ebbs and flows of artificial-intelligence research. In short, the original artificial-intelligence research was mostly based on rules—people summed up various rules and input them into the computer, which the computer itself was not able to do. This advanced approach, a machine-learning technique, was based on statistics that allowed computers to find the most probable and appropriate models from large amounts of data and multiple paths.

In the last couple of years, artificial intelligence has become vibrant again, thanks to the upgraded version of machine-learning technology, a deep-learning approach based on a multilayer computer chip neural network. With a multilayer chip connection, the technology replicates the mesh-connection mode of a large number of neurons in the human brain, supplemented by exquisite reward and punishment algorithm design and big data, training the computer to efficiently search for models and rules from the data, thus opening up a new era of machine intelligence.

A few people with determination have saved the excitement for the return of artificial intelligence. In China, Baidu was one of the first companies to deploy artificial intelligence. It seems to have done a lot of things that other companies had not heard of before. About seven years ago, Qi Lu and I talked about the tremendous progress in deep learning in the United States and were determined to make a big move into the field. Finally, in January 2013, I officially announced the establishment of IDL (Deep Learning Institute) at the Baidu Annual Meeting, the first research institute under "Deep Learning" in the global business industry. I was the dean, not just because my knowledge of deep learning was best but also because my name would show the great importance I attach to deep learning and summon scientists who have stuck to this field for many years.

This is the first time Baidu ever set up a research institute. Our engineers are researchers, and researches have always been closely integrated with practical applications. I believe that deep learning will have a huge impact in many areas in near future, although some of those areas are not in Baidu's business scope. Therefore, it is necessary to create a special space to attract talent, and let workers try various innovations freely, do research in areas unfamiliar to Baidu, and explore the revolutionary path of artificial intelligence for all mankind.

"Smart" Being Upgraded

If the enlightenment phase of artificial intelligence can be called the 1.0 era, then obviously now it has entered the 2.0 era. Machine translation is a typical case. In the past, machine-translation methods were based on the rules of words and grammar. Humans constantly summed up the grammar rules for the machine, but they couldn't keep up with the changes in human language, especially the context. So, machine translation always made mistakes such as translating "怎么是你" into "How are you?" ("怎么是你" means "It's you" with an astonished tone in Chinese.)

Later, SMT (statistical machine translation) appeared, with its basic idea to find out common vocabulary combination rules by statistical analysis of a large number of parallel corpora (collections of written material) and trying to avoid strange phrase combinations. SMT already has the basic functions of machine learning. There are two stages of training and decoding: the training stage is to let the computer construct a statistical translation model through data statistics and then use this model for translation; the decoding stage is to get the best translation by using estimated parameters and given optimization objectives.

SMT research has been in the industry for more than twenty years. For phrases or shorter sentences, the translation works well, but for long sentences, especially for languages with different structure, such as Chinese and English, the result is far from satisfactory. Recently, the NMT (neural machine translation) approach has emerged. The core of NMT is a deep neural network with countless nodes (neurons). After a sentence of a language is vectorized, it is transmitted in layers in the network and transformed into a form that the computer can "understand." Then it undergoes multiple layers of complex conduction operation. A translation in another language is thus generated.

But the premise to apply this model requires a large amount of data; otherwise, it is useless. Search engines like Baidu and Google can discover and collect massive human translations from the Internet and feed such huge data to the NMT system, which can then train and debug a more accurate translation mechanism better than SMT. The more Chinese-English bilingual corpus we store, the better outcome NMT will achieve.

SMT used all local information before, and the processing unit was the phrases, or segmentations of the sentence. At the end of decoding, the translations of several phrases were stitched together, without fully utilizing the global information. NMT uses global information, first encoding the information of the entire sentence (similar to human reading through the entire sentence before translating) and then generating a translation, based on the encoded information. This is the advantage and the reason why it is better in terms of fluency.

For example, a very important step of translation is word-order adjustment. In Chinese, we put all the attributives in front of the central words, while in English we put the prepositional phrases behind the central words that they modify, and machines often confuse this order. The advantage of NMT in word-order learning brings the fluency of its translation, especially in the translation of long sentences.

Traditional translation methods are not completely useless, and each method has its own advantages. Taking idiom translation as an example, there are often customary translations, free translation instead of literal translation. Idioms must be translated in the corpus with corresponding contents. Nowadays the needs of Internet users are diverse, and translation involves many areas, such as speaking, résumés, and news. So it is quite difficult to meet all the requirements with just one method. Therefore, Baidu has been combining traditional methods, such as rule-based, instance-based, and statistical-based methods, with NMT to advance research.

In this machine translation model, humans do not need to look for voluminous language rules by themselves but need to set mathematical methods, debug parameters, and help computer networks to find their own rules. As long as humans enter a language, machines will output another, without worrying about what has been done in the process. This is called end-to-end translation. It sounds amazing, but in fact, Bayesian methods and hidden Markov models in probability theory can both be used to solve this problem.

Taking the Bayesian method in information distribution as an example, we can construct a personality-feature model described by probability. For instance, one of the characteristics of male-reader model is that the probability of clicking on military news is 40 percent, while for female-reader model it is 4 percent. Once a reader clicks on military news, the gender probability of the reader can be back-deduced, and with other behavior data, this reader's gender and other features can be judged after comprehensive calculation. This is the magic of mathematics. Of course, the mathematical methods used by computer neural networks are much more than those described in this example.

The idea of artificial-intelligence technology methods like machine translation dictates that the amount of data must be large enough. The Internet provides a massive amount of data that scientists used to dream of but found it hard to realize. The original intention of the birth of the Internet was to facilitate information communication, resulting in information explosion, which promoted the development of artificial-intelligence technology.

Take chess as an example. In 1952, Arthur Samuel developed a checker program with similar competitiveness as amateur masters. The rules of checkers are relatively simple, and computers have far more advantages than humans, but chess is much more difficult. When Zhang Ya-Qin, former president of Baidu, was the dean of the research institute at Microsoft, he invited Xu Feng Xiong, a computer talent in Taiwan who developed the famous Deep Blue chess robot at IBM (International Business Machines Corporation). During the 1990s, Deep Blue was the most qualified representation of artificial intelligence, concentrating "wisdom" on one supercomputer, using multiple CPUs (central processing units) and parallel computing technology; it continuously defeated human chess masters and finally defeated international chess champion Garry Kasparov in 1997. But shortly after the game, IBM announced that Deep Blue retired. Zhang Ya-Qin said to Xu Feng Xiong, "You should invent a Go robot and come back to me when it can defeat me," but Xu Feng Xiong did not come to him, even after he left Microsoft.

Some bottlenecks of Deep Blue are hard to overcome: although it can handle the calculations on the chess board, it becomes powerless facing the variability of a different order of magnitude on the Go board. Based on the decision-tree algorithm, the model of exhausting all possibilities is beyond the carrying capacity of computers. Although the algorithm is continuously optimized, still it cannot break the computational barrier. The Oriental wisdom represented by Go seems to be inviolable at the level of artificial intelligence, but a new era is coming.

Internet Congress

The computer intelligence represented by Deep Blue seems to have nothing to do with the Internet. However, the development of cloud computing and big data has finally brought artificial intelligence and the Internet together.

The combination of these two complementary powers allows us to acquire a different wisdom model from the Deep Blue era. Multichip distributed computing, coupled with the big data accumulated by humans, and with the new algorithm beyond the decision-tree as a chain, embodies a perfect combination of human intelligence and machine intelligence.

From 2016 to 2017, AlphaGo (a Go robot) was insurmountable in the Go field. Its "thinking" is different from humans and Deep Blue. In short, it contains the data of millions of humans' Go games. For a more professional interpretation, it can be said that the Monte Carlo search algorithm and deep-learning-based pattern recognition contributed to AlphaGo's achievements, the most important of which is deep learning, which its predecessor Deep Blue does not possess.

According to the research of all parties, AlphaGo does not think about how to play but learns the game of master players (big data). It records every situation in all the games, trains millions of situations as input, and then predicts the next step for human masters through a multilayer neural network. Through ingenious neural-network design and training, this multilayer neural network models the "sense" of master players for the current situation; the winning percentage in the previous games is already known. When actually playing, the computer can record the game by visual recognition, compare it with the previous game data to find the same mode (situation), search for different situations for further development, and choose some high-quality possible steps for the next move according to the winning percentage from the past game histories, instead of trying each possible step. That greatly reduces the amount of system calculations and saves the system from "exhaustion." It is like a human being; it doesn't exhaust all the possible points but picks some points based on experience and feeling. Humans still have to calculate and compare which points are better after selecting a few; for the machine, this calculation is handed over to the Monte Carlo search algorithm.

A vivid but not necessarily accurate metaphor would help to explain. Monte Carlo tree search is an optimization of previous decision-tree algorithms, which even after a high-quality possible step is decided still has to exhaust possibilities for the next choice, branching at each choice, resulting in an exponential explosion on the number of optional paths.

The Monte Carlo method shows the subtlety of probability. Suppose that under a certain Go situation, the deep-learning network gives three choices of a move: A, B, and C. Taking these three points as the root nodes, we can imagine three actual trees, each having countless branches. A Monte Carlo search does not exhaust all branches, but sends three million ants to start from A, B and C, one million for each point; the ants quickly climb up to the treetops (that is, the black and white take steps alternately until one wins; generally that will be within two hundred steps). There must be some ants reaching the highest point (that is, the outcome is determined). If the ant goes to the end, the black wins; if it does not, the white wins.

Suppose three hundred thousand among the one million ants starting from point A reached the end point, with five hundred thousand from point B and four hundred thousand from point C having reached the end point. The system would conclude that choosing point B for the black will bring a higher winning percentage and then take point B as the next step. This is the probability sampling algorithm, which greatly reduces the amount of computation compared to the item-by-item exhaustion method.

Why send one million ants instead of one hundred thousand or ten million? This is usually based on the computing power of the computer, and a rough estimate of the competitor. If a higher winning percentage can be reached by sending one hundred thousand ants, then we do so. The more ants sent at the same time, the higher the computing power required.

The CPU chip and the GPU (graphics processing unit) chip simultaneously perform neural-network calculations and Monte Carlo tree searches to simulate a massive final situation, which is incomparable for human computing power. Since deep-learning models master players, it seems that artificial intelligence has the big picture of human beings, in the data of millions of games between master players.

I believe smart readers, even if they don't know much about mathematics, have basically understood how AlphaGo operates, although the specific algorithms and strategies are far more complicated than the aforementioned description. AlphaGo showed us the current level of artificial-intelligence and deep-learning technology development. However, there are many institutions and individuals who do similar research and development, and like the eight immortals crossing the sea, each show special prowess.

Once recorded by the Internet in the form of data, human behavior becomes an endless treasure that nourishes artificial intelligence of all types and helps humans. Machine translation, speech recognition, and image recognition are all based on a large amount of data provided by the Internet, as well as user click behavior. Why is the accuracy of the Baidu search engine unmatched by other search engines in China? Baidu has the largest amount of data, the most advanced algorithm, and the most profound accumulation. Every click of the user is actually training the Baidu Brain behind the search engine, telling it which information is most important for the users.

When artificial intelligence encounters setback, people start believing that it is difficult for a machine to think like a human being, but this also shows opportunity. After the 1990s, people realized that artificial intelligence does not have to think in the same way as humans do, as long as it can solve humans' problems. So, when the linguist Noam Chomsky was asked, "Can the machine think?" he quoted the Danish computer scientist Edsger Dijkstra: "Can the submarine swim?" The submarine cannot swim like a fish or a person, but it possesses outstanding underwater ability.

When we look back at history, not just the development of the Internet, we can find the entire human industrial development gestating artificial intelligence. Kevin Kelly said that the self-reciprocating motion of the steam-engine piston is a delicate design and such a self-response already contains the element of "evolution": the pursuit of automation is the evolutionary dynamic of artificial intelligence.

For example, at the beginning of the industrial revolution, steam engines first appeared in coal mines and pits. Since early steam engines were inefficient and energy intensive, they could only be used where coal was particularly abundant and cheap. When coal was mined, a lot of water was produced, which needed to be extracted from the mine. With this demand and enough cheap energy, the idea of the steam engine was generated. Once it went into service, the steam engine constantly developed, eventually pushing the industrial revolution. Artificial intelligence is similar. When you have enough data, the data is the fuel on which the artificial-intelligence engine can run.

Thanks goes to the development of the Internet and the data records generated by all human activities, without which the computer would lack the means of learning. Thanks also goes to those artificial-intelligence explorers who are not all computer scientists. Some of them do biological research, some do engineering research, some study the automatic iterative optimization of mathematics and computer programs, and some update the collaborative architecture of computer chips. The various research results have merged into the sea and finally become today's artificial intelligence.

Magnates Contending for the Market

The media's astonishment regarding AlphaGo in 2016 was actually a delayed reaction. Back in 2007, Geoffrey Hinton, a giant in the field of artificial intelligence, had noticed that "the gale is raging before the storm is about to burst."

At that time, one of Hinton's students, with the help of Google big data, applied Hinton's earlier research findings to speech-recognition technology and achieved remarkable success. Hinton couldn't help but sigh: "The past failure was only due to the lack of data and computing power."

In the second decade of the twenty-first century, everything is ready for artificial intelligence, and a time of fierce competition is just about to begin. Since 2015, the artificial-intelligence entrepreneurship has continued to percolate. According to the data analysis of the artificial-intelligence industry released by CB Insights, a US venture-capital data organization, artificial-intelligence investment exceeded $1 billion in the first quarter of 2016, and there were 121 investments in the second quarter, compared with twenty-one investments in the same period of 2011. From the second quarter of 2011 to that of 2016, the amount of investment in artificial intelligence exceeded $7.5 billion, of which more than $6 billion was generated after 2014.

Wuzhen Index: Global Artificial Intelligence Development Report shows that in the first two quarters of 2016, more than sixty artificial-intelligence start-ups were set in China, with an investment amount of $600 million. In the past year (2015), 202 investments were made in mainland China in the field of artificial intelligence, involving a total of $1 billion (about 6.8 billion yuan), and the market is huge.

In 2016, Academician Tieniu Tan, vice president of the Chinese Academy of Sciences and vice chairman of the China Artificial Intelligence Society, said that the global artificial intelligence market in 2015 was $127 billion. In

2016, it was estimated to reach $165 billion. By 2018, this figure will exceed $200 billion.

China, the United States, and Britain are the most important development regions for artificial intelligence. The United States is the origin of the Internet and artificial intelligence. It has a unique talent advantage, coupled with a strong technology base and huge research funding; it remains number one in this field. In addition to Google, Facebook, Microsoft, Amazon, IBM, Apple, and many other giants who made large investments in the field of artificial intelligence, there are nearly one hundred large and small companies focusing on AI business. For example, x.ai, which specializes in natural language processing, attracted a three-round financing of $340 million.

The United Kingdom's long-established famous universities keep gathering talent in the field of artificial intelligence under the circumstance of shrinking manufacturing industries. DeepMind Technologies, which developed AlphaGo, has benefited from the people who graduated from UK universities.

Amazon launched Alexa Smart Voice Assistant and Echo Smart Speaker to compete in the voice market with Apple, Google, and Microsoft. In June 2016, Amazon's CEO Jeff Bezos revealed in an interview with US technology blogger Walt Mossberg that Amazon's investment in key projects in the field of artificial intelligence has lasted for four years. "The project team has more than 1,000 people, and what you see is only the tip of the iceberg."

In September 2016, Microsoft announced the establishment of a new artificial intelligence R&D group under the leadership of Executive Vice President Harry Schum. He led thousands of computer scientists and engineers to integrate artificial intelligence into the company's products, including Bing, Digital Assistant Cortana, and robot projects. At the end of the year, Microsoft officially released a service that can develop chat robots and announced that it will provide CPU service for the OpenAI artificial intelligence lab cofounded by Elon Musk and Sam Altman, former president of the startup incubator Y Combinator.

Facebook also has its own artificial-intelligence lab and a team like Google Brain—Applied Machine Learning. The mission of the group is to promote artificial-intelligence technology in a variety of Facebook products. In the words of Mike Schroepfer, the company's chief technology officer, "About one-fifth of Facebook's engineers are now using machine-learning technology."

Of course, AlphaGo's owner, Google (which acquired the program from DeepMind Technologies), is not content with the ability of playing Go, and its artificial-intelligence investment has been expanding over the past years. In 2012, Google only had two in-depth learning projects, but this number exceeded one thousand at the end of 2016. At present, Google utilizes deep learning in its search engine, Android, Gmail (free webmail service), translation, maps, YouTube (video website), and even unmanned vehicles.

China has a huge amount of business-application scenarios, users, and data, as well as the largest group of talent; it has made rapid progress. In addition to BAT (short for the three major Internet companies in China: Baidu, Alibaba, and Tencent), Huawei Technologies, and other giants who intensively developed artificial intelligence, many other artificial-intelligence companies in vertical industries are also emerging. In various Internet forums in 2016, heads of Internet companies in the fields of e-commerce, social media, and search engines were all leading the topic of artificial intelligence, reporting large or small achievements.

In 2016, Baidu's speech-recognition accuracy rate reached 97 percent, and face-recognition accuracy reached 99.7 percent. Baidu's platforms TianSuan, TianXiang, TianGong, and Tianzhi, through the cloudization of Baidu Brain, have successively opened up the technology and capabilities of Baidu Brain to the whole society.

A Convergence of Super Brains

A few defenders in the field of machine learning more than a decade ago are now the most valuable talent. After the rise of artificial intelligence, available talent became one of the two scarcest resources in the open-source world, in addition to data.

The professional knowledge behind artificial intelligence is highly correlated with basic disciplines, such as mathematics and biology. Artificial-intelligence scientists are the best in these fields, which is especially rare. However, in China, there are less than two hundred doctoral and postgraduates students in this field per year, which is far from the demand of numerous new startups. This is the case abroad too. In 2015, Uber headhunted 40 of the 140 researchers at Carnegie Mellon Robotics Academy, causing an uproar in the industry.

The foregoing is not all about talent competition. AI companies are basically sensitive to the problem of brain drain. In 2017 and 2018, many academic stars quit their previous position or started a business, which highlights reality in the burgeoning AI field. Those who do well academically must be able to realize their potential.

Baidu is the main representative of China's artificial-intelligence industry. A large number of top workers have joined Baidu: Wang Haifeng worked at Microsoft before joining Baidu; Andrew Ng came to Baidu from the United States; Zhang Ya-Qin came to Baidu from Microsoft; and Lin Yuanqing left NEC American Lab, which is full of machine-learning experts. Jing Kun, the creator of robot Xiaoice, came to Baidu from Microsoft; Qi Lu, the highest-ranking Chinese executive who was in a US giant technology company and the authority of artificial-intelligence technology, gave up the position of Microsoft vice president to join Baidu. At the same time, many skilled workers started in Baidu before creating their own artificial-intelligence application company. Baidu is the epitome of China's vitality in attracting and cultivating artificial-intelligence talent.

Many super brains in the industry have come together to create an epoch-making Chinese brain. We have experienced the PC era, are now in the era of mobile Internet, and are about to enter in the era of super intelligence of the Internet of Everything. A "super brain" environment will become possible, for which we process the gathered various kinds of data, and Baidu is creating such an environment. The purpose is to let artificial intelligence penetrate the lives of Chinese and even the world, like water and electricity, and to turn everything in the world toward the direction of "informatization."[2] For example, Baidu Brain has its own eyes, ears, mouth, and cognitive decision-making ability. Overall, it is equivalent to a child, but its local abilities such as translation, speech recognition, and image recognition greatly surpass that of human beings. We open these abilities to everyone to develop and explore various artificial-intelligence

---

2 *Informatization* is a term put forward by Kevin Kelly in *The Inevitable: Understanding the 12 Technological Forces that Will Shape Our Future*. It means that software devours everything, and everything is informatized. Even a table will be able to upload its own data, such as sales history, frequency of use, etc.

applications. Baidu Brain has become the tool of many developers and the operating system of artificial intelligence. It promotes the formation of artificial-intelligence standardization, comprehensively serving companies, entrepreneurs, and individuals.

Therefore, we are eagerly calling for the establishment of China Brain, a national platform for deep-learning servers, algorithms, and application infrastructure. It will be a manifestation of the all-around upgrading of China's competitiveness and a powerful accelerator for the Chinese renaissance.

Technology: An Extension of Human Life

Relative to human data nourishing artificial intelligence, I need to discuss our users, the countless consumers who support Baidu and high-tech Internet development.

Today, in addition to the influence of large companies such as Google, Microsoft, and the BAT group, the decentralization of the Internet and big data technologies enables small businesses, talented technicians, and even users to change the situation.

> The secret to AOL's success is actually simple: discover what people want most and then offer it to them. Its success also benefited from strong marketing propaganda; it established a user-friendly corporate image. President Jim Kimsey played a key role here. He knows nothing about computer technology, but from the perspective of a business person, he sets his feet deeply rooted in the public. "The center of my world are consumers; we want to be the Coca-Cola of the Internet world."

I emphasized the importance of users in *Battle in Silicon Valley*. In the eyes of engineers, "user" is strictly and rationally defined; the user requirements-development-feedback description is rigorous in the technical documentation. But the development of the Internet not only facilitates technical services but also provides a stage for thoughts and emotions. We can say that the Internet has created a kind of opinion-based user.

Many of our programmers and engineers enjoy Baidu's relaxed environment for technical workers, which is simple and reliable. Some technicians are not good at communication or complicated interactions; they are eager to develop a wide range of products. Users' various emotions are different from the engineer's. Engineers in the laboratory may not experience the type of events of ordinary people, nor the complex and volatile transactions and emotions. People from the media and public relations can better understand users' emotions. Staff from our public relations department sometimes complain about technicians not knowing users' psychology; when encountering problems, they often think that modifying code is the only thing that needs to be done. But human emotions cannot be fixed, which troubled us deeply. One of the problems we must solve is how to break the gap between technicians, merchants, and ordinary users. We need inspired product ideas and humility to allow cross-border learning.

The thinking about users' daily needs is an ongoing task that needs to be sustained. But as far as the theme of this book is concerned, we are engineers, after all, and shall never forget to meet the needs of users with technology and numbers. We use technology to accurately differentiate data and serve different users.

The trend of digitization has been discussed in books like Nicholas Negroponte's *Being Digital* and Kevin Kelly's *Out of Control: The New Biology of Machines, Social Systems, and the Economic World* and *What Technology Wants,* and it also always exists in the technician's mind. We are surrounded by living data. Data often sounds dangerous. For example, will private data be sold? We will continue to talk about this topic later. Here, in simple terms, the data in artificial intelligence is by no means data like an ID-card number or a password. Today's artificial intelligence focuses on finding the overall model of chaotic data, thereby optimizing production and service. Advances in translation, speech recognition, and image recognition are the best examples. These chaotic data will be of great value for human beings after AI sorts out their regular patterns, from daily speech recognition to credit-fraud prevention in the financial field to antiterrorism security at the national level.

Technology should adapt to users all the time. The product side responds directly to the user's needs and should continuously optimize the performance of the technology. We believe that good artificial intelligence should be unobtrusive, unlike a power source with unstable voltage. It should continuously improve accuracy and optimize product details. For example, some companies are good at speech-recognition technology, but their overall design of the input method is not convenient enough, which affects users' experience. Baidu has unsuccessful product examples that need to be changed with the help of users.

Data and technology are not cold; they will show humanness when combined with good artificial-intelligence methods.

There was an active degree of migration between Dongguan and other cities in China after the antipornography campaign in Dongguan during early 2014. A senior news editor told us that he thought the map figure Baidu created with visualization technology transcended the news event itself, and there was a feeling of overlooking the world as a god. The Baidu Migration Index reflects human migration through data-visualization techniques. The migration of people in the digital age is only a small page in the epic of human migration for millions of years, but, it is a historic page in the era of big data.

I think the map signifies a historic moment in the era of artificial intelligence, as smart-map technology illustrates human activity. Artificial intelligence doesn't have humanity, but when combined with the developer's creativity and ideas, it can provide a new perspective, even a different type of human compassion.

Both computers and the Internet are part of the artificial-intelligence body. Each data set is a record of human activity and humanity. Artificial intelligence thus finally emerges like the soul. It can be humane.

Data Avenue

One philosopher said that human beings are a kind of ongoing existence. Baidu has accumulated a large amount of map data, supplemented by the designer's wisdom and various sophisticated algorithms, that can depict various human mobile behaviors and observe people's living conditions on the road.

The maximum amount of Baidu Map's daily service reached 72 billion recorded events, each one representing a record of human activity. It can also be seen from the map data that cities in Central and Western China are increasingly connected. The bustling traffic heat map and rhythm are like the city's life pulse. The eye of the map has a large field of vision.

People of our generation have all heard the song of Tong Ange: "In order to live, people are running around, but they are staggered in fate." I hope that with the help of artificial intelligence, human trajectories are not only staggered but also intersected and merged into rivers and lived endlessly.

A young scientist at Baidu Big Data Lab who majored in biology studied the laws of fish-school movement at Princeton before he returned to China. Upon seeing the migration map, he said that Baidu let him know that human data can also be studied like a fish school, only with more convenient methods, so he decided to join Baidu. In 2016, he and his colleagues used the search data changes on Baidu Map to accurately predict the decline in iPhone sales. With the help of data, the Big Data Lab provides intelligent awareness for a variety of urban life and business operations.

In 2014, the Ministry of Transport proposed to deepen reform in the following ways: promote pragmatic innovation; accelerate development of four modes of transportation; accelerate the construction of a market- and enterprise-oriented industrial technical-innovation system that combines production, education, and research; and promote the transformation of scientific and technological achievements into transportation productivity. The Ministry of Transport made efforts to establish a multichannel and multimodal travel-information service system and initially establish an integrated traffic-information service platform to release real-time travel information to the society, solving problems such as poor travel information.

In this context, Baidu proposed the China Smart Transportation Cloud Service Platform Cooperation Plan and jointly established a cooperation platform with the Research Institute of Highway of the Ministry of Transport and the National Intelligent Transport Systems Center of Engineering and Technology. Relying on the Ministry of Transport's key technology project, Research and Demonstration on Open Public Travel Information Service Based on Cloud Computing, this plan activates existing data, establishes a province-wide information-resource sharing and exchanging mechanism, promotes the sharing of service information between government and enterprises, and is open to the whole society.

The smart map can report the degree of road congestion according to the user's moving speed and can also intelligently avoid the odd and even license-number restricted routes. Using virtual-reality technology, we can find our way on the map as we do on the road. Based on traffic big data, coupled with algorithm assistance, and responding to the needs of traffic-management departments, intelligent map systems have been able to provide solutions for the mitigation of urban traffic congestion, greatly reducing the pressure from the traffic control department.

Smart map's collection of geographic data has made many smart projects possible. High-definition map technology with centimeter-level accuracy has been applied to the development of unmanned vehicles. At the World Internet Conference in 2016, an unmanned Baidu vehicle was publicly tested and commissioned in Wuzhen. The total mileage was about 3.16 kilometers; the vehicle passed three traffic lights and made one U-turn, with pedestrians, cars, and electromobiles on the road and complex weather condition of rain, mist, and haze. The test competes with the road tests conducted by Silicon Valley peers in North America. This is a small step for unmanned vehicles, but it will definitely be a big step in artificial intelligence.

Artificial intelligence does not grow on trees; it is a natural result of the progress on computer-network technology and data-processing technology over the past decades and the data set of human beings. The intelligent development of Baidu Search and Baidu Map is a microcosm of this process.

AI Is Neither a Myth nor a Joke

There are plenty of news reports about robots in various mass media; some of them are just for fun. For example, news broke the other day about a robot attacking people in an exhibition. The truth is that an educational assistant robot fell from the stage and hit someone. Another news item concerns a cemetery equipping itself with robots to embolden grave guards, but such robots are actually just toys, more like a spoof. If we look back at history scientifically, we will find that artificial intelligence is neither a myth nor a joke, but a real product from the labor of mankind. We don't need to fear or worship it.

In the field of artificial intelligence, a scientist's description of technology is often straightforward and modest. Jun Wu, a former Google engineer, said in 2003 that he and his companions worked together to improve keyword search accuracy of Google. One of the main problems they solved is which meaning of synonyms to search to fulfill the user's demand. For users, if the search results are inaccurate, they will continue to search for a synonym or select a result that is not ranked among the top search results. At this time, users are actually making a keyword collocation, and the system will record the collocation relationship; now the goal is to return the result faster and more accurately. Wu said:

> The way we did it may sound less technical. We have sorted out the keyword colloca-tion that users have searched for years, stopped one of the company's five largest data centers during the four-day vacation of the Independence Day of the United States in 2003, and used these four days to specialize in the matching of each keyword. This is actually an exhaustive method. It is to solidify the word combination relationship that users often choose. The next time users take a similar search, the system can present the result faster and more accurately.

In fact, the technical logic in the field of machine translation is similar to the tactical exhaustive method applied in the search Wu described. According to the *New York Times*, at a meeting of Google's translation department on a Wednesday in June 2016, an article by Baidu in a core journal on machine translation provoked the staff's discussion. Mike Schuster's words restored silence to the conference room: "Yes, Baidu has published a new paper. It feels like someone has seen through what we have done—papers share simi-lar structures and results." Baidu's BLEU score (bilingual evaluation under-study, a measure of the accuracy between machine translation and pure human translation) basically matched Google's achievements in internal tests in February and March. Quoc V. Le didn't feel uncomfortable. He concluded that this is a sign that Google is on the right track. "This system is very similar to ours," he said quietly.

Quoc V. Le is a PhD student of Andrew Ng. He may not know that the paper has nothing to do with Ng but was independently finished by the natural language department. The *New York Times* report about Chinese companies is of course simplified. However, Ng believes that Chinese media should not always think that foreign technology is more developed. The media tend to report the post-knowledge as a breakthrough. In fact, many advanced inventions in the field of artificial intelligence were done by Chinese people, but Chinese media may neglect them at first; then when foreign technology comes up with the same inventions, they regard the latter as breakthroughs.

Baidu has released an NMT-based translation system one year ahead of schedule, and Google then launched a similar system in 2016. We can conclude that the basic skills of the most cutting-edge explorers in this field are similar. In the end, he who possesses the most profound accumulation of knowledge and best optimization will win.

Today's artificial intelligence is different from the past; it has changed thinking rules into data and strategy. In the past, humans always wanted to design the perfect logic for computers and constantly abstracted human logic rules into functions and then entered them into computers. Today's artificial intelligence is based primarily on advances in big data foundations and algorithms. In other words, the explosion of artificial intelligence today is based on the Internet outbreak in the late 1990s. With the help of the Internet, significant data will be generated continuously. We should notice that the data here is not what users consciously put in, such as name, age, address, hobbies, etc., but the data automatically generated when they use the Internet. For example, each search and each click is a kind of data, and each moving trajectory (such as your walking or driving routes) also counts.

China is already the world's number one manufacturing power; now it needs to upgrade its "soft power." Spirit and culture are soft power, as are calculations and data. The combination of this kind of soft power with the traditional industry makes "smart+." It will be practically integrated into our production and life visibly and tangibly.

Why Must Baidu Do It?

Instead of asking what Baidu wants to do, we'd better ask, "Why must Baidu do it?"

Every company has its own strategy and tactics. In 2013, the Chinese mobile Internet entrepreneurial trend began, and many companies invested huge sums of money in this huge bottomless pit, showing their brave strategy. Baidu focuses on the long-term and scientific aspects. At that time, few people noticed that Baidu was working on artificial intelligence. Today, artificial intelligence has become famous throughout the world, and some people are amazed with the advance and firmness of Baidu's strategy. Baidu recognizes the nature of the Internet information industry in advance; once determined, it resolutely takes its own path and does not care about an outsider's judgment. Baidu's multiparty layout and key breakthroughs have helped to set up the foundation of artificial intelligence when the whole world started to pay attention to it.

We didn't direct Baidu's artificial-intelligence technology toward activities such as playing Go or predicting the results of singers' competitions. Instead, we are focused on developing internal strengths and concentrated on transforming artificial intelligence into practical services that can improve human life. We not only apply deep learning to a few areas such as speech recognition, machine translation, and street-view house-number recognition but also apply its success to significantly enhancing users' experience.

In 2013, Baidu Navigation was the first in China to announce that navigation would be permanently free and brought China into the era of free navigation. Now, we are opening the data interface of Baidu Maps for others to develop and use. Everyone can use the positioning technology and solutions provided by Baidu Maps to save a lot of cost compared to the traditional GPS tracker. Delivery companies can use this platform to plan the optimal delivery route, and game developers can develop location games like Pokémon Go. We open Baidu Brain so that more people can use artificial eyes and ears to serve themselves. We open up the deep-learning development platform PaddlePaddle so that more people who are interested can create their own artificial-intelligence services. We also hope that nontechnical people can learn to use data intelligence to optimize their work, improve their personality, and pursue their own ideals.

A lot of college-entrance-examination candidates must have used the Duer personal-assistant robot to help them choose their college and major. In China, each field attracts many participants. When I was a student, people called the college-entrance examination "a thousand troops and horses crossing the wooden bridge." Similar to the map data, Duer analyzes the huge data of college-entrance examinations, responds to and senses a student's desire and anxiety through deep-learning technology, and tries to provide an accurate response. Here, artificial intelligence records not the map trajectory in the physical space but the spiritual trajectory of a student's growth.

In the early 1990s, I moved to the United States to study computer technology. At that time, there were many young people like me, with the desire to change the world with code, traveling between China and the United States like migratory birds. If there were a data map at the time to record these transoceanic trajectories, it would be very interesting. Now artificial-intelligence scientists have brought the fire to China again, and I believe that the flame will burn even more warmly because China has enough fuel. China's population of educated people is large, and the popularity of computers and mobile devices is high. A large amount of data makes China uniquely advantageous in developing and applying deep-learning technologies. With this advantage, we are ready to create the legend like the Silicon Valley in the 1990s.

What Baidu has created is not only the frontier development but also data infrastructure and a deep-learning and development platform, gathering the wisdom of people.

Before Trump was elected as the President of the United States, more than a hundred Silicon Valley elites issued an open letter saying Trump's election would be a disaster for innovation. This shook me a lot. If the innovation in the United States is really affected, then who will take the banner to lead the direction of innovation? Can we move the world's innovation center from Silicon Valley to China?

Talented individuals are indeed coming to join us. Baidu also set up a laboratory in Silicon Valley to get closer to American skilled workers. The China Brain plan proposed by Baidu is the equivalent to any super project. Seventy years ago, top scientists returned to China from abroad to build great projects with great enthusiasm. Will such glorious achievements reappear today?

Of course, it must be noted that the great projects of that era were contingent on national investment and industrial policies. After the end of the Cold War, the country's competitive pressures had decreased, and investment in cutting-edge technology had also been greatly reduced. Musk's rocket-development project was actually the result of the government's decision to transfer the NASA (National Aeronautics and Space Administration) rocket technology and team to him. The Chinese government has strong determination and investment, and the development of the artificial-intelligence industry is a consensus. It is the best of times; it is the most uncertain of times. Artificial intelligence is a way to adapt to uncertainty. Large and small companies invest in artificial-intelligence research and development, bringing competition and diversification, which should form a benign interaction and growth.

With regard to the uncertainty, the White House report is already exploring the impact of artificial intelligence on employment. The rapid development of Silicon Valley and the decline of the central manufacturing industry have increased the country's rift. Some people have enjoyed progress and others have been left behind. Baidu wants to be the gatherer of talent, and Chinese companies must strive to build an ark to take the masses to the intelligent era.

Dr. Wang Haifeng, Chief Technical Officer of Baidu, was elected as an ACL (Association for Computational Linguistics) Fellow in November 2016. The youngest fellow of ACL, he is also the first Chinese to be the chairman in the association's fifty-year history. The selection committee wrote in a comment to him, "Wang Haifeng has made outstanding achievements in the fields of machine translation, natural language processing, and search engine technology, both in academia and business industry, and has made tremendous contributions to the development of ACL in Asia." At the beginning of 2017, Qi Lu, a well-known scientist and executive in the field of artificial intelligence, joined Baidu. These developments indicate the trend of international talent mobility. Thousands of outstanding artificial-intelligence scientists in China are working together to create the future of mankind.

The Future Has Come: Anxiety and Dreams

Not long ago, Amazon's no-cashier supermarket astonished shopaholics. Behind this special shopping experience, there is the shadow of the cashiers being laid off. Today, with various online customer services being replaced by machine customer service, shorthand translation being replaced by speech recognition, and even cashiers, drivers, factory workers, copy clerks, and lawyers being replaced by artificial intelligence, how shall we face the world? What kind of support should the government and enterprises provide for workers? How should we adjust the economic and social structure to adapt to the era of artificial intelligence? We want to listen to the needs of ordinary people. This is also the original intention of this book by our artificial-intelligence team.

Peter Thiel is a venture-capitalist genius in the Silicon Valley, equally famous as Marc Andreessen. He is the creator of Mosaic and Netscape, two of the first widely used web browsers, so he is good at grasping technical trends and capturing dark horses. In 2016, he became famous again because of his accurate prediction of Trump being elected President of the United

States. He said in 2011, "We wanted flying cars; instead, we got 140 characters." The 140-character Twitter was once very popular, but Peter Thiel clearly saw what was missing from the bustling Internet. He criticized people for slowing down the pace of progress, insisted that hippie culture replaced progressivism, and said venture capitalists are keen to invest in light-asset companies, most of which are mobile Internet companies, such as Airbnb and Uber, but without clear planning and confidence for the future. He believes that in the Internet+ era, humans have made great progress at the bit level and little at the atomic level. Therefore, he decisively invested in rockets, anticancer drugs, and artificial intelligence.

I agree that mobile Internet entrepreneurship has covered up the progress we pursue. Baidu must strive for its own direction and contribute to the advancement of human core capabilities. Thiel said that Americans in the early twentieth century were willing to try new things and dare to plan and implement the decades-long moon-landing plan. However, today people do not have such a plan. Only the venture capitalists are looking around for the added value and timely pleasure. Baidu is willing to imagine an intelligent world and realize it. It wants to make artificial intelligence a new operating system, not only for computers but also for the world. At the same time, it seriously contemplates and responds to the challenges of artificial intelligence in advance, eventually making a different world. So, I said that we must make it!

The intelligent revolution is a benign revolution in production and lifestyle and also a revolution in our way of thinking. Huge opportunities and challenges coexist. Subsequently, I will discuss the specific aspects of the intelligent revolution; detail the breakthroughs made on deep learning, such as visual recognition, speech recognition, and natural-language processing; and depict the upcoming intelligent society along dimensions of manufacturing upgrading, unmanned driving, financial innovation, management revolution, and smart life. Furthermore, I will explore how people should cope with the development of artificial intelligence, and I will take the pulse of the intelligent revolution.

2

THE HISTORICAL MISSION OF ARTIFICIAL INTELLIGENCE: LET HUMANS KNOW MORE, DO MORE, AND BE MORE

Lu's Conjecture

The 1980s were idealistic and unforgettable. At that time, many Chinese people pursued scientific progress and were hungry for knowledge, especially college students, who were crazy about reading. Scientists like Chen Jingrun[3] have become idols of many people.

In 1987 on the campus of Fudan University was a thin young man wearing glasses who carried a big backpack every day. He was very energetic and fond of thinking about mysterious questions. His classmates all called him Chen Jingrun.

When he graduated, he wrote a farewell note in the yearbook:

I would like to present my latest research results and bid farewell to my school friends.

Lu's conjecture: HI=>C⊃HB

(where H: Human   I: Intellectualized

C: Computer   B: Brain)

---

3   Chen Jingrun is known throughout the world for his proof of Goldbach's conjecture of the "1+2" theorem. Xu Chi's reportage, *Goldbach Conjecture*, with Chen Jingrun as the protagonist, was published in the first issue of *People's Literature* in January 1978 and became known to many households at that time.

The idea is that humans will eventually make the computer intelligent and far better than the human brain. You can learn from Chen Jingrun; perhaps this pearl on the crown of computer science is you.

Lu is Qi Lu, former president of Baidu. Lu's conjecture is certainly not a scientific theorem like Chen Jingrun's conjecture, nor is it a joke. Viewed from today, it is more like an advanced epiphany. But, why was he so confident to make such a prediction thirty years ago?

At that time, Qi Lu only had some sly feelings:

> The computer has brought us to extraordinary knowledge and experience. At that time, we wrote the chess program when we were studying in [the] computer department. Although it was very simple, I got an intuition that just if given enough time, computers can certainly be smarter than humans in the future. I had such an intuition at that time, so I wrote those words. Shortly afterwards, I had a precious opportunity to study computer science in Carnegie Mellon University.

When he graduated from Carnegie Mellon University, Lu Qi wrote, "Know more, Do more." Later, he added "Be more." He felt that "Be more" was most important, but he didn't realize this in the first place. "Be more" can also be translated as "become more." People are on the road to enrichment.

The organic combination of the three needs has pushed human progress. The history of mankind is to constantly discover new things. The more we know, the more we can do. The more we do, the more we can experience, and the more enriched our life will be. When we experience more, we will know more. This is a positive cycle and the main theme of human progress.

Computers make us know more, do more, and be more, and artificial intelligence is the latest consequence of the striving. From this perspective, we can have a clearer insight into where artificial intelligence comes from, where it goes, what its nature and standards are, and what people and organizations who are interested in artificial intelligence should do.

The Next Wave

We are witnessing an era of computers and digital information, a long-lasting and necessary tide of human history. And artificial intelligence is the driving force to push the tide to the next high point. It will have a lasting revolutionary impact on our society. Such influences involve economic and social aspects, such as industry and technology. But, in the end, the artificial-intelligence revolution will allow us all to move forward in a completely different way and write a new history.

First, we need to understand the essence of human progress. We are passionate about things that allow us to know more, achieve more, and gain more experience.

The continuous development of computing power is following the direction of human progress and becoming the essential manifestation of it. Especially after the emergence of computer programming, humans have begun to progress in an unprecedented speed. The core model includes the following key steps: humans capture various phenomena in the universe and gain experience especially through deliberate observation; then through calculation, the information is effectively organized, processed, and refined, allowing humans to have a deeper and more abstract understanding of a certain phenomenon, which forms knowledge; humans act by using the knowledge and interacting with phenomena, finally achieving the progress.

While the IT (information technology) industry is creating information worth tens of billions of dollars, it is enabling humans to make great strides with three core dimensions of information organization (helping humans know more), task completion (helping humans achieve more), and experience enrichment (helping humans gain more experience).

As a new upgrade of human computing power, artificial intelligence is still pushing human progress through those three dimensions. Moreover, since artificial intelligence is a revolutionary, higher-level intelligent computing system (ICS), its impetus for human progress is unprecedented and revolutionary.

The main structure of the modern digital computing system is determined by the organization of resources. The essence of artificial-intelligence computing, in simple terms, is very different from von Neumann's control flow structure, which uses linear memory and Boolean functions as baseline calculation operations. The new paradigm is neural-network computing, which is characterized by distributed representation and activation patterns. Here, variables are represented by vectors superimposed on shared physical resources (such as neurons) and are calculated by activation of neurons. The network's topology and activation modes provide a huge computational space to capture rich knowledge efficiently and naturally (through the hyperparameters, weights, and activation functions of topology). Compared to localized representations in the von Neumann architecture (where variables are represented by dedicated or localized physical resources such as registers) and symbolic calculations, neural-network computing is more natural and powerful in learning and representing the rich semantic knowledge of the physical world and society.

With the help of neural-network computing, the next wave of artificial-intelligence technology can enhance the current computing system in two dimensions:

- Automatic layering feature learning. This is a substantial increase in machine-learning capacity; the key to today's machine learning is feature engineering. For example, Baidu Brain already has tera-scale parameters, hundreds of billions of samples and character training.
- Advanced cognition, especially perception. This is a huge catalyst for next-generation devices such as unmanned vehicles and next-generation platforms such as natural-language conversations.

The powerful artificial-intelligence computing will help to generate many new varieties of intelligent systems, such as machine lawyers, machine analysts, medical robots, and intelligent customer-service staff.

Another developmental direction of artificial-intelligence computing is to focus on the organization of various systems that serve specific physical architectures and physical elements, which incudes the home, offices, factories, and the like. The basic model is that through the use of various raw signals of IoT (Internet of things) sensors, the "sense" system of artificial intelligence recognizes and perceives the physical architecture; the "cognitive" system needs to organize information and learn more about the physical architecture to predict, judge, and make decisions, so as to make all kinds of physical systems more intelligent.

At present in the scientific research field, artificial-intelligence computing can provide more advanced modeling capabilities and become a catalyst for multidomain and new-wave scientific research.

Artificial intelligence can provide an additional opportunity to create an integrated business computing system (BCS) platform for business organizations, for functions such as recording business objects, like system-design models and transaction records, and business processes, like ERP (enterprise resource planning), CRM (customer relationship management), or system design and human work-activity imitation, such as communication, collaboration, reading, writing, information seeking, etc.

At present, the sensing systems of artificial intelligence have broader and newer business opportunities. On one hand, more subsystems of a sensing system can be built and deployed for physical environments or physical systems, such as assembly lines, factories, etc. This will provide more information tools and greater automation for labor-intensive manufacturing and business services in the future. On the other hand, the rapid advancement of natural-language processing allows us to scan and analyze text documents and information and to extract a variety of high-value professional knowledge, while building and deploying a keen text-understanding subsystem that can yield a lot of high-value knowledge and business returns.

The maturity of the artificial intelligence cognitive system represents a longer-term future of the intelligent era, and all industries, occupations, social systems, and lifestyles will be reshaped. If the digital society can be summarized as "information at the fingertips," then the essence of the artificial-intelligence era can be summarized as "knowledge is everywhere; intelligent interaction is everywhere." This wave is undoubtedly a huge opportunity for most people.

For business organizations, the good news is they will have many opportunities to upgrade, transform, and enter new growth areas; the bad news is when the big wave comes, no one can stand still; if you don't seize the opportunity to move forward, you will fall behind and even be abandoned by the times.

For entrepreneurs, just look at the list of great market opportunities, and you will find many opportunities to create future business giants. The historical opportunity of intelligent startup that highlights entrepreneurship is just around the corner. At the same time, new business leaders of the intelligent age will also be born.

For investors, the chance has to come to find a seed company that belongs to the smart age, support it to grow into a future corporate empire, and finally realize a huge return.

For the country or the government, every huge technological revolution is always accompanied by the rise and fall of the country's destiny. Some countries or governments can seize the opportunity given by history, leap to the top, and win a prosperous national development for a long period. National or government-level decision-making, policy foresight, strategic investment, and scientific and effective route implementation are all crucial ingredients to seize such historical opportunities.

Of course, as more and more people perceive the wave of artificial intelligence, more and more entrepreneurs will be engaged. It is foreseeable that there will be confusion at the beginning.

In Silicon Valley, everyone said that they will invest in AI+X in the future. For example, there is an investment company that favors artificial intelligence—Bloomberg L.P. Michael Bloomberg, former New York City mayor, is its CEO. After investigating hundreds of companies, it got confused because all startups claim to be concerned with artificial intelligence. For the investment company, the first question is how to identify companies so as to make a targeted investment. This prompted us to start thinking about what the standards are for artificial intelligence companies. Which are real artificial intelligence companies, and which are not?

Realistic Criteria for Artificial Intelligence

In the history of mankind, the emergence of every new technology inevitably will be accompanied by various discussions, reflections, and even tit-for-tat debates. Faced with the gradual rise of a global and revolutionary technology wave of artificial intelligence, people have diverse excitement, doubts, and concerns. Some of them are more emotional, such as whether artificial intelligence will replace humans. This is actually comparing artificial intelligence with natural intelligence.

There is a variety of research and literature about natural intelligence, including the suggestion that the internal mechanism of the human brain is actually quantum computing. For artificial intelligence, there is currently no accepted definition. At this stage, there is no need to pursue a more correct standard definition. We may discuss pragmatically what kind of intelligent system the current technology allows us to make.

Two types of computing systems are regarded as artificial intelligence:

- The first type is basically equivalent to the subsystem framework of the intelligent computing system (ICS). It takes data as input, extracts information from the data, and builds models that turn some of the phenomena we are concerned with into knowledge. We call this type of artificial intelligence system *general AI* and define *general intelligence* as the ability of a machine to acquire knowledge and achieve certain goals.
- The second type refers to cognitive ability, which is found in human beings. It can perceive ("see," "listen," and "feel"), reason, plan more and more contents, and control the sensorimotor movement. We call this type of artificial intelligence system *cognitive AI*, which means a machine with the ability to sense, reason, plan, and use sensorimotor control.

Another dichotomy of artificial intelligence systems is *narrow AI* and *strong AI*. Strong AI is a system that uses the same algorithm to solve a large scope of problems. In principle, the strong AI system can learn and adapt to solve new problems without human involvement. The narrow AI system uses specific algorithms to solve specific problems, such as chess, image identification, and so on.

In summary, Table 2-1 is an overview of the state of the artificial intelligence system and presents a pragmatic and feasible definition.

Table 2-1: Status of Artificial Intelligence Systems

Definition of Artificial Intelligence	Narrow AI	Strong AI
General-purpose Intelligent System	Intelligent Customer Service	Future Search Engine, Baidu
Cognitive Intelligence System	Unmanned Vehicle	DeepMind Company

Intelligent computing systems are directly related to big data. Each data set has its origins and systems behind it, which is why data is generated, and the core of the data is knowledge.

The general-purpose artificial intelligence system has the fundamental capability to extract information from the data by using algorithms and computing systems. Once we have the information, we can do a lot of things. We can predict, realize automation, and solve any problem we want. Because information tells us what people and society need, we can find the answer. Therefore, the first level of artificial-intelligence development is general-purpose artificial intelligence.

The breakthroughs in deep learning in recent years have been mainly at the perceptual level, especially visual and speech recognition, as well as natural-language understanding. But this is only the beginning. The next step is at the cognitive level because perception only turns the external world into a symbol that can be recognized by the system through light, the vibration of the sound, or the communication of the language. The most important point is to understand its meaning. When the system sees a picture, it knows what kind of objects or people are on it and what they are doing.

Almost all companies in the field of artificial intelligence can now be placed in four quadrants. What most companies do is actually narrow AI, which solves only one problem or solves one or two narrow problems. Playing Go, playing cards, or driving a car are all narrow AI. However, strong AI uses one system to solve all problems, which is similar to human intelligence. Strong AI is the long-term goal of artificial-intelligence development, and its real implementation will take at least two or three decades.

Now, Baidu, Google, Microsoft, and Facebook are working toward strong AI. The criteria of artificial-intelligence ability, or whether a technology is truly artificial intelligence, is still based on whether humans can know more, do more, and be more with the help of it.

For example, Baidu's many technical analyses based on massive search data have been almost impossible for human labor. Now, through artificial-intelligence computing technology, we have obtained much unprecedented knowledge, so humans know more and can do a lot of unprecedented judgments for more "impossible" achievements. The representative examples of unmanned technology and natural-language interaction technology have both gradually changed movement and sensation. In the past, humans used their eyes to see and ears to hear. In the future, we will see without eyes and hear without ears, and humans will gradually have new ways of sensing and experiencing a new world.

Therefore, whether a company is really doing artificial intelligence can be measured from the foregoing perspective: Which of the four quadrants does the technology belong to? Does it allow humans and machines to know more, do more, and experience more?

Many companies in the United States and China say they are artificial-intelligence companies. Some say cloud computing is artificial intelligence, and some say big data is—but these are only two parts of the artificial-intelligence system. Ultimately, artificial intelligence is judged by big data, cloud computing, algorithms, training time, total investment, and the comprehensive strength of software and hardware.

These are not achieved overnight, nor can they be generalized. People play different roles on the road, which is full of thorns and obstacles, and choose different stops. Each person and enterprise will have different achievements. Some have just started, while some have yielded much fruit.

Baidu Brain can be regarded as a typical example of the comprehensive strength of artificial intelligence. The breakdown of its capabilities enables us to better understand the entry threshold and basic standards for the artificial-intelligence industry. If a company that claims to be involved in artificial intelligence does not have the following capabilities, then we can only conclude that it is not ready to enter the artificial-intelligence field.

Baidu Brain is a combination of hardware foundation, data foundation, and algorithm capabilities. It is the trinity: cloud computing, big data, and artificial intelligence, the core of Baidu's technology strategy. With cloud computing being the infrastructure, big data as the fuel, and artificial intelligence as the engine, they jointly promote the physicalization of the Internet, sending the Internet technology and business model of the digital world back to the physical world and changing society in an all-around way.

Cloud computing is the most fundamental and physical part of Baidu Brain. It is IaaS (infrastructure as a service). Baidu Brain's superior computing power comes from here, the army of high-performance computing hardware. This army has hundreds of thousands of servers and is managed by an advanced cluster operating system. It is called an artificial-intelligence supercomputer.

For deep learning and training, Baidu independently developed GPU and FPGA (field programmable gate array) heterogeneous computing servers, which can be extended to 64 GPU/FPGA cards, increasing the density of traditional servers by 16 times. One server can complete the training of 100 billion data models. Baidu pioneered the development of an FPGA-based artificial-intelligence processor, providing 10TOPS computing performance. The computing efficiency is sixty times greater than the mainstream twenty-core server; it can achieve four to eight times the performance of ordinary servers in artificial-intelligence and big-data applications.

Baidu's advantage is not only the excellence of a single machine but also the outstanding system and a strong overall work capability, integrated with the finest individual machines. For the intelligent scheduling and resource management system of the GPU cluster, the pooling management and dynamic scheduling of computing, storage, and network resources can be realized, and the overall efficiency and average utilization rate of the computing cluster reaches up to 80 percent. Using heterogeneous hardware for online products, the user's request latency is reduced by four-fifths, and the computational efficiency is increased by more than tenfold.

This system covers the largest GPU/FPGA cluster in China (thanks to new chip technology), the largest Hadoop/Spark cluster (new concurrent data-processing technology), and the most efficient data center, with new heterogeneous computing technology, whole-cabinet server technology, 100G RDMA (remote direct memory access) communication technology, and operation/maintenance technology, providing the computing power needed to develop artificial intelligence with full control.

The fuel tank is also full. Through the years of service to large-scale businesses, such as search and video technology, Baidu has accumulated a large amount of data: trillions of web data, billions of search data, tens of billions of video, image and voice data, tens of billions of positioning data, etc. Data is the fuel of artificial-intelligence algorithms and another basic condition for the development of artificial intelligence.

Excellent algorithms and models can add the fuel to hardware. Baidu has attracted the world's top scientists and engineers and continues to innovate in theory and practice. It has built the world's largest deep neural network, supporting trillions of parameters, hundreds of billions of samples, and hundreds of billions of features training. The number of neural network layers is far beyond one hundred.

The combination of hardware power, data fuel, and algorithmic soul makes Baidu's PaaS (platform as a service). The uniqueness of Baidu's PaaS is that artificial intelligence serves as a horizontal service throughout the platform. Through deep learning and machine-learning techniques, combined with superior computing, massive data, and excellent algorithms, we have outstanding capabilities in speech, image, natural-language processing, etc. That creates unique knowledge maps, user portraits, and business logic, open fully to users. Users can easily use a variety of algorithm modules, development tools, and data engines to serve their own business purposes. We named the different platforms as TianSuan, TianXiang, and TianGong, which are targeting three areas: intelligent big data, intelligent multimedia, and intelligent Internet of things, respectively.

At the top level of SaaS (software as a service), Baidu's artificial intelligence easily condenses into many vertical industry solutions that penetrate all walks of life. But we are also pursuing smart industry environments with our partners, such as an education cloud, financial cloud, traffic cloud, logistics cloud, and so on. We believe that the ability to build an intelligent industry environment is also an important criterion for evaluating artificial intelligence.

Another important criterion on hardware, data, and algorithms is the corporate culture of artificial-intelligence enterprises, that is, the "soft power." Search technology is the pioneer of artificial intelligence and the gateway to the world of Internet digitization. Its developmental process and technology core have laid a foundation for future artificial intelligence. First, search engines must deal with large-scale data. Second, search engines must be equipped with large-scale machine learning at the same time. It is impossible to do it manually due to the large quantity of data. Finally, and most fundamentally, the developmental process and engineering-development culture of search engines must be consistent with the development of artificial-intelligence systems. They are all based on data. They bring value to users by extracting features and patterns. People become collaborative in the search process, and the professional capabilities and work habits thus formed are very suitable toward the development of artificial-intelligence business; they are accumulated as the culture of artificial-intelligence enterprises together with the massive data. Therefore, Qi Lu's practice at Microsoft is to train talented individuals by working in Bing. Staff who have worked in Bing can handle any other department, and those technologies seem to be very simple compared to searching. This culture is certainly not perfect, but like the neural network, it can be continuously developed and perfected under the guidance of the correct method.

Artificial Intelligence+ World

Just like Internet+ in previous years, people are now discussing artificial intelligence+, adding business, industry, medical care, education, etc., behind the "+." From the perspective of "know more, do more, be more," artificial intelligence is bringing fundamental change to the world, so it is an issue of "artificial intelligence+ world."

First of all, the intellectual revolution will have a profound impact on everyone's daily life. As a simple example, the breakthrough development of artificial intelligence will promote a closer interaction with computing devices.

Previous human-computer interactions were carried out through only the mouse and keyboard. Microsoft has reached today's scale through the innovation of human-computer interaction through the mouse, keyboard, and GUI (graphical user interface). The biggest contribution of Apple and Steve Jobs to the world was to change the world by using fingers to interact with the GUI on the screen. The changes in the era of artificial intelligence are even greater, and humans will be able to communicate with any device in natural language.

Natural language is the most effective and common form of communication. People use language to communicate, as it is the most natural and most widely used way. The realization of natural-language interaction between human and machine means humans do not need to understand every application or how to use every product; instead, they just operate them. Future cars will directly communicate with us, and houses will also talk to us easily.

We have already seen the prototype of this kind of intelligent interaction, such as some intelligent-assistant systems. In the United States, people turn the house into an intelligent system through Amazon's smart speakers. In China, Baidu's Duer team has also made a lot of cutting-edge exploration in this field, and we have the opportunity to completely change the way of human-computer interaction in daily life.

Artificial intelligence will greatly accelerate the speed of human innovation and the efficiency of social-value production. The scale of society will be changed, and it will be completely different from the past. Every revolutionary process of human progress begins with the discovery of new knowledge, which works for the radio and also for the Internet. The discovery pattern of future knowledge will undergo fundamental changes. In the past, humans pondered and discovered the rules of the real world. After the arrival of the digital world, with the help of artificial-intelligence data-processing methods, people and machines will discover new knowledge together. This means the speed that humans create new businesses and social processes and change will increase; the world will be completely new. The process of "know more, do more, be more" has been greatly accelerated.

Finally, artificial intelligence will bring about a new industrial revolution. Why do many people think that after industry 4.0, humans will enter a new stage of the digital society? It is because intelligent systems will be able to extract data in the real world, capture knowledge, and thus better help humans to perceive the real world. It will also profoundly and widely change the real world from economic, social, and cultural levels. We are indeed in a very exciting time, very similar to the early industrial revolution, but artificial intelligence will have a broader and greater impact on society than the industrial revolution.

The traditional manufacturing industry is principally based on equipment, electrical appliances, and electricity. Its production pipeline is basically built with large-scale investment, and it is difficult to adjust afterward. It costs much money and time for an automobile manufacturer to rebuild its production line. When the data intelligence, automation, and precision predictions are completed in the manufacturing industry, the industry will be completely renewed. Future manufacturing processes will be modular, and all digitally controlled. When an automobile manufacturer wants to manufacture another type of car, it will no longer need to rebuild the production line but only will need to transfer the new product module's application programming interface (API). This will completely change the manufacturing base as well as the manufacturing efficiency.

The core of this change is data and knowledge, which means all the steps of manufacturing—the process, technics, design—will be controlled by data.

Another example is the pharmaceutical industry. In the past, the birth of a new drug required a long-term research-and-development process to find an effective treatment for a certain disease. In the future, with the help of artificial-intelligence computing technology, we can quickly discover the rules and find personalized genetic drugs under the analysis of the combination of huge genetic data with massive health information.

From the national level, artificial intelligence is not only the catalyzer of overall competitiveness, but also a godsend opportunity to transcend other countries. China is a big manufacturing power, and the sheer volume of data is unmatched, which means that we have the opportunity to extract much more knowledge than others. When we know more and see more, we can do more and become stronger than others. In the era of artificial intelligence, more knowledge can help us to become invincible in national and industrial competitions. In terms of manufacturing alone, if China can seize this opportunity and complete a truly intelligent upgrade, then there is no way for other countries to compete with us. However, the way to achieve such intelligent manufacturing requires an overall strategic plan.

Currently, the United States and China are the two countries that pay most attention to artificial intelligence. People tend to compare Baidu and Google, which can also be seen as a microcosm of the comparison between China and the United States. I think the two companies share lots of similarities. The similar origins have yielded much in common in the company culture. The advantages of Baidu in China are similar to those that Google has in the United States.

There are also differences. The space and speed in some areas of Baidu's innovation may be bigger and faster than Google, thanks to China's different national conditions.

The degree of innovation in the mobile Internet in China has surpassed the United States on many aspects. For example, Baidu app (a mobile-phone application) innovates in information flow and can achieve refined searching with the core technology of artificial intelligence. Baidu has more opportunities for mobile-Internet innovation in China than Google in the United States, because the IT industries in the two countries are different. Baidu Finance can rely on the advantages of the Chinese market and data to make a revolutionary improvement in the financial industry with artificial-intelligence technology. But in the United States, the financial industry is relatively stable, and it is difficult for Google to get involved in it.

Another example is unmanned vehicles. Now, Google and Baidu are taking the lead, with Google being slightly ahead of Baidu, but the future is uncertain. Because China has many auto manufacturers and a more coopera-tive environment, working with artificial-intelligence companies like Baidu may create many opportunities for innovation, and the speed of innovation will be faster. In contrast, in the United States, automakers are concentrated in Detroit, and it is difficult for artificial-intelligence companies to seek cooperation with them.

In short, China and the United States face smart upgrade needs in the industries of unmanned vehicles, finance, medical care, and overall manufac-turing. However, China's macro environment gives a smart enterprise such as Baidu more opportunities and space than Google has gained in the United States.

So, what kind of responsibility should Baidu take in this global wave?

In the United States, the IT industry system generally relies on five com-panies: Apple, Google, Facebook, Amazon, and Microsoft. The supporting enterprises are led by the five enterprises.

In the context of the artificial-intelligence era, Baidu is a supporting enterprise, and we must strive to make greater contributions toward the intelligent revolution in China and the world with such an opportunity. Specifically, we must strategically position ourselves with energy. First of all, Baidu is a Chinese company, and Baidu Brain should be a pathfinder and founder. Baidu's intelligent cloud is provided to all industries and will pro-mote, empower, and drive them.

As the forerunner of Chinese artificial intelligence, Baidu has innovated and explored in multiple dimensions and gradually formed the prototype of its own intelligent environment. For example, medicine and education are areas that have a large potential of artificial-intelligence application because the essence of the two fields is data. The professional quality of senior teach-ers and old doctors comes from their experience (data). In the future, machines will automatically analyze the data and assist doctors to prescribe medicine and assist teachers to individualize teaching. Medicine and health care can make people live healthier, and education can give people more knowledge. Therefore, artificial intelligence has great value.

In addition, the unmanned vehicle will also be realized by perception, cognition, and knowledge acquisition. At present, it takes some time for the unmanned vehicle to be commercialized, but after successful commercializa-tion, the entire society will be greatly transformed. This is not just about cars and transportation, with unmanned vehicles connecting to the Internet on their own and being guided by it, and a lot will take place in different industries.

The practical scope of artificial intelligence is wide, with a great opportunity to completely transform everything. Of course, the implementation of the strategy should be carried out step by step, with a firm direction and steady pace.

Challenge on Enterprises: How to Put into Practice

In terms of work attitude, Qi Lu said, "Head above cloud, feet on ground," which means that the head must be above the clouds to see far and clearly, but the feet must step on the solid earth step by step.

The first obstacle of artificial intelligence is practice. Artificial intelligence means a huge change, and it will take a long time. First, we need to identify a beneficial and practical user experience. Second, the scenario must be clear; actual users should experience values through smart assistants, unmanned vehicles, and information. Lastly, a business model is essential for sustainability.

So, the important challenge is whether we can find a scenario to put user experience and user value into practice. Then we need to find the right business model and build an innovative cycle, such as data→knowledge→user experience→new data. Once such a cycle is set, then the artificial-intelligence business can roll forward like a snowball.

Most important, CEOs must pay attention to artificial intelligence, which is the start. Then they must invest a certain amount of resources, including hiring people who really understand artificial intelligence and who can help make decisions. The company may specialize in the retail industry, manufacturing industry, or tourism industry. Regardless, it must develop an effective and intelligent strategy according to its own business status and then implement it resolutely. It is necessary to give the executor sufficient power to implement intelligence into specific business through effective strategic planning.

We may use the flywheel model to break down the implementation steps of the artificial-intelligence strategy.

THE HISTORICAL MISSION OF ARTIFICIAL INTELLIGENCE

First of all, keeping the direction of the artificial-intelligence development in mind, organizations must reorganize the positioning of the enterprise, set new development directions, determine the new mission, and then imagine opportunities that enterprises should seize in the artificial-intelligence era.

Second, organizations must develop an intelligent strategic blueprint based on the company's new positioning. This requires the corporate leadership to position the company's vision in the upcoming wave of artificial intelligence, making trade-offs between which fields to enter and which fields to quit.

Certain principles must be followed when making "in and out" decisions. Silicon Valley consultant Geoffrey Moore's hierarchical framework is a good example for the evaluation of the artificial-intelligence wave. The key point is to enter high-growth categories and jump out of low-growth categories. The wave of artificial intelligence will create new categories with huge growth potential, such as unmanned vehicles, robots, echo equipment, and dialogue systems. At the same time, artificial intelligence can also hamper certain industries, because new artificial-intelligence-driven products will replace the previous products in some way—for example, the new artificial-intelligence hardware and software stack will damage the investment based on the old HW (hardware) stack. A good practice is to develop a complete list of new high-growth categories, regrowth categories, and headwind categories so that the leadership can make systematic and principled decisions.

The next step is to decide at the beginning whether the product is valuable and incomparable. It should be emphasized that in the era of artificial intelligence, the key to uniqueness is whether products possess unique data assets (which bring unique knowledge).

Next is to understand uncertainty, risk/reward, and time frames for prediction and management. The horizon model is a good framework in this regard for making decisions and organizing portfolios. The general approach is as follows: during the H1 period (the first eighteen months) focus on current core business. Dedicate the H2 period (the next eighteen to thirty-six months) to investment, in the creation of a profit engine. The H3 period (the next thirty-six months plus) should be committed to long-term investment in greater potential yet higher risks. The wave of artificial intelligence provides a very rich opportunity for H2 and H3 periods, and some artificial-intelligence investments can even help to bring improvements in the H1 period. In general, artificial intelligence is at a very early stage, with many unknown constraints and uncertainties. To truly understand artificial intelligence, it is important to make decisions according to principle and practicality.

During the implementation stage of enterprise's artificial-intelligence strategy, we must first adhere to the principle of "structural integrity"; that is, to be consistent in product experience, technical architecture, and business model. If we are moving toward ICS (server/client architecture) or investing in an autonomous system, then technical decisions need to be synchronized with product and business decisions.

Secondly, it is essential for enterprises to keep up with the technological roadmap of artificial intelligence, as well as the current rapid development of deep-learning technology.

For the leading companies in the artificial-intelligence industry, they need a strong vision to change the globe, world-class technology foresight, strong research teams, and research plans, which are all consistent with the presentation of our enterprise vision, intelligence technology, and product development. DeepMind, Google, Baidu, and some aggressive pioneering companies are all doing this.

At this stage, updating the research mechanism is also an important step. Traditionally, the IT industry and academia are not good at commercializing research results. The recent Otherlab and OpenAI and a few other artificial-intelligence startups are actively recruiting research teams, which is a new trend. Various organizations (universities, large corporations, and training and R&D institutions) need to work together on relevant research to develop structured and sustainable solutions.

Investment is an important factor that companies need to consider. With the deepening of the intelligent revolution, the battle for talent has escalated, resulting in a continuous increase in the cost of artificial-intelligence development. Due to huge long-term return on investment (high risk/high return), some startups are able to raise large amounts of money. The key behind developing an investment plan is to prioritize resources and a well-thought-out decision-making process that reflects the risks of artificial intelligence.

When all the objective conditions are ready, people become the decisive factor. Leadership skill is a far-reaching and rare element. Since the artificial-intelligence wave is based on a completely different core technology (with neuroscience as the core), it requires advanced management capabilities from top management teams. At the same time, AI-driven emerging industries are so diverse and interdisciplinary (from genetics to robotics, whatever you can imagine) that companies need to hire an innovative person. (But it's not easy because today's social life is very specialized in many areas.) Bill Buxton, principal investigator at Microsoft Research, built an innovative team for senior management.

The core of the artificial-intelligence innovation flywheel is the data→knowledge→user experience→new data feedback cycle. Optimizing the capacity and speed of this feedback cycle is a very important part of planning.

Finally, I want to emphasize that the core of the strategy is to actively set goals based on the current status quo and inference, and to implement actions that can achieve the goals.

What Kind of Macro Environment Do We Need?

A good macro environment is indispensable for enterprises and scientific-research institutions, just as the China Brain plan is a call for the intelligent infrastructure at the national level. To usher in the era of artificial intelligence, the government also needs to create suitable soil through macro planning.

First, make sure that the data is accessed easily. Data is increasingly a strategic asset for many organizations and can be seen as a new type of natural resource. The government can obtain data through policy development and open it to the whole society in order to stimulate more innovation.

Second, use open-source tools and platforms. The artificial-intelligence wave requires a new silicon and software stack, for open-source tools and platforms, like PaddlePaddle in the early days; it is very important for them to be used by developers and innovators. Looking ahead, we need to continuously reduce barriers for participation and use more tools and more modules systematically. Just as AWS (Amazon Web Services, which provides on-demand cloud computing) makes computing easier, some AI-as-a-service can also make artificial-intelligence technology more accessible.

Third, innovators can quickly nurture the market conditions and policy systems of products for users, which is also very important because the innovative flywheel needs a fast feedback cycle.

Fourth, encourage sustainable applied research. In the early stages of the artificial-intelligence wave, sustainable application research, especially its development, can acquire knowledge from data. ML (machine learning) algorithms create intelligent experiences at the core of this revolutionary flywheel.

Fifth, these points are bound to lead to the issue of talent. It is a key factor to educate and train more talented individuals who can design and implement machine-learning algorithms and become data scientists.

Finally, through the new structured approach, the information and knowledge of the public becomes organized and accessible, which is critical for the innovation of artificial intelligence in many enterprises.

The Culture and Long-term Management of a Smart Society

It will take decades for this wave of artificial intelligence to fully play its role. Ambitious and world-changing predictions call for long-enough investment. Therefore, purposeful long-term management plays an important role in the wave of artificial intelligence. It will revolutionize the business and management culture throughout society.

Specifically, the government needs to establish and acquire a larger "license envelope" that allows the management team to have a longer time span to nurture big forecasts, which has increasingly become an important part of the senior leadership's function and power. Musk said, "If innovation fails, they should not be punished." For companies that are affected by the wave of artificial intelligence and need to enter a new pattern, the renewal and transformation of the entire organization is critical, and the senior leadership team needs to face and manage such revolution.

One factor associated with long-term management is to create and nurture new organizational structures to adapt to the changes brought by artificial intelligence. Alphabet (the name of Google's reorganized umbrella company) is one of the earliest attempts. In this regard, Chinese companies have done more management innovation than the United States.

In addition, culture is an enduring force of an organization that can last over generations of leadership and business activities. For many mature companies (such as Google and Baidu), entering the wave of artificial intelligence is a major challenge—they always need to acquire new talent and new technology patents and to create a new and consistent culture. Being proactive, patient, and persistent is paramount because cultural transformation is one of the most challenging jobs for a mature company. In addition, forgetting the old-mode working methods is more difficult than learning the new AI methods.

As we are in the early stages of the development of artificial intelligence, recruiting and maintaining experts in this field is very important to managers.

In general, long-term management is a key to capture all important opportunities, including but not just limited to the wave of artificial intelligence. How to adjust the structure to attract more capital and talent and put more emphasis on changing the game pattern is an unusual and challenging job for business leaders. Facing deep, more interesting, and more challenging issues is a sign of human progress.

Current State of AI Technology Development

New research and articles on artificial intelligence and deep learning are published every day. Like the Renaissance, all the sciences are rapidly changing because the essence of science is to observe the world and summarize knowledge. We are now more and more capable of observing the world. Once we choose the observation angle, then we can use the algorithm of deep learning, and soon new knowledge will come out.

Nowadays, every scientific field is careening forward, including but not limited to physics, biology, and materials science. So, in general, humans are in a state of rapid development. Let us put our minds in the clouds again, with quantum computing ending this chapter.

We note a very interesting connection between artificial intelligence and the neural computing framework: both are based on the distributed representations of large vectors whose basic operations are linear algebra rather than Boolean algebra. It tells us that human brains operate in a similar way as matter does. Some scientists have even suggested that the human brain operates similar to the principles of quantum physics, quantum-computing theory, and quantum-computing algorithms. Regarding the combination of quantum computing and artificial intelligence, we find that Microsoft and Google have already established quantum artificial-intelligence laboratories, and China will also launch projects in this field. We should not hesitate over the issue of whether we have quantum computing, but instead discuss the time to have it. We must have it. After people have far-reaching ideas about this, early quantum machines will exist maybe five years later, or even within five years.

Why is quantum computing becoming so important? It has an essential relationship with artificial intelligence. The core of quantum computing is superposition, which is a change in the state of power when two or more quantum states are added together. Current computer digits, 0 and 1, have only one state at a time; unlike that, quantum superposition can have four states at the same time, and the computing power increases exponentially.

One of the main advantages of quantum computing is that it can solve many data problems. For example, data encryption and decryption are both created with prime numbers, which makes the process very difficult. To encrypt a big number by general algorithm may take forever, but with a quantum algorithm, it can be done very quickly. Thus, it will be natural to use quantum algorithms for machine learning.

Related to this, the hardware must also be innovated. The current hardware is mainly based on Boolean algebra, but the core computation of deep learning is the computation of matrix and tensor, not the calculation of 0 and 1, and must be differentiated. Quantum computing is also exactly the same. When each quantum changes its energy level, it is a computational of matrix and tensor. Nature actually computes in this way, and so does the human brain. Scientists such as Matthew Fisher, Pan Jianwei, and Qingshi Zhu believe that the essence of consciousness is quantum entanglement.

In 2007, *Nature* published the results of a laboratory led by Graham Fleming of the University of California, Berkeley, which uses femtosecond laser technology to deliver photosynthesis complexes in a very short period. When the laser is irradiated, the light echoes on the composite like drums, which means that the energy of the photons is not transmitted to the reaction center through a single path, but through all possible paths by quantum coherence. It proved that the quantum effect plays an irreplaceable role in chlorophyll photosynthesis. This drum-like quantum echo is the mapping of nature and the light of human wisdom. The discovery of quantum effects in living organisms has greatly inspired human's new exploration of quantum computing and human-computer integration.

Even though quantum computers have not yet been implemented, many people are already thinking about how to do machine learning with quantum computers. There are already many frontier articles and research results in this field. Assuming that quantum computers are invented ten years from now, it will bring a fundamental change to the artificial-intelligence industry, because quantum computing is completely consistent with the core calculations of artificial intelligence and deep learning. We are actually taking a detour now: all algorithms are changing themselves into Boolean algebra, using 0 and 1, to simulate a differential equation.

The scale and energy of quantum computing and DNA (deoxyribonucleic acid) computing will far exceed today's silicon-based computing. As engineering technology advances, we will have a new computing experience (such as quantum chemistry and quantum materials).

Quantum computing will be widely used. The first implementation may be in traditional agriculture. As the photosynthesis example shows, we know that plants are also calculating. In the future, crops may be calculated and designed with the help of computers. Therefore, quantum computing may bring a super change for the whole society, and it is very likely that human civilization will be completely digitized.

In short, quantum computing is not just metaphysics but exactly the future of human progress as stated in "know more, do more, be more." In this respect, we need as much imagination as possible. We must dare to imagine, while standing on solid ground. Thirty years ago, Qi Lu wrote on the graduation album, "The pearl on the crown of the computer science is you." If our generation can't achieve it, the next generation will. This is the unchanging dream of "capitalized humans."

3

ARTIFICIAL INTELLIGENCE SUCCEEDING IN BIG DATA AND DEEP LEARNING

Change in the Cycle of History

What are we talking about with regard to data?

For most people, the word *data* may mean figures on the monthly water, electricity, and heat bills, the red and green index on the K chart of stock, or the bunch of obscure source codes in a computer file.

From the standpoint of artificial intelligence, the meaning of data is far more extensive. Data spans human civilization, from the initial sounds, words, pictures, and figures, to every image, speech sound, and video of the electronic age; every mouse click in the Internet age; every finger slip on a mobile phone; every heartbeat and breath; and even all human actions and trajectories in economic production.

Nowadays, human beings are able to turn various things, big or small, into data records and make them part of their lives, including both the eternal gravitational wave and the complex and subtle DNA. Data has been immersed in every detail of our lives. Just as biologists believe that half of human tissues are made up of microbes, in the digital age, half of our lives are already data.

History always spirals forward. Let us go back to the past, long before the birth of artificial intelligence, when human beings also practiced the discovery, calculation, and utilization of data.

More than five thousand years ago, ancient Egyptians summed up the law by observing the position of the astrological signs: the Nile River began to flood when the wolf star (Sirius) appeared on the eastern horizon every year. They worked out plans for agricultural farming according to this and summarized the cycle to determine the solar calendar for a 365-day year. The distant Sirius has no causal relationship with the earth; when it appeared in that position, the earth moved into a certain solar term. This is the predecessor of the correlation calculation in the era of big data.

More than four thousand years ago, Stonehenge appeared on the territory of today's Britain, large stones each weighing about fifty tons, forming a circular array. This is an original timepiece. When summer solstice comes, its main axis, the ancient road leading to the stone pillar, and the first sunshine in the morning will be on the same line; in the opposite direction, the last rays of the winter solstice will also cross the stone gate. The ancients used a cumbersome stone gauge for data measurement. This was the earliest data-visualization technology.

More than two thousand years ago, Claudius Ptolemy studied the motion of celestial bodies and established the three laws that laid the foundation for astronomy. His method is very interesting; in a nutshell, he conveyed a right idea in a wrong way. Originally, he mistakenly thought that the trajectory of celestial motion was circular, when it is in fact elliptical. In order to describe the actual motion curve of the celestial body with a circle function, he used multiple circular nested motions—as many as forty. That's equivalent to fitting a total function with multiple circular-motion functions. His work is the earliest idea of fitting functions.

What is a fitting function? When there is a lot of data, we can imagine the data as a lot of points in a coordinate system. How do you find a function that allows its curve to cross as many points as possible? If these points are distributed regularly, such as in a linear distribution, they can be described by linear equations.

If the distribution points form a parabolic shape, then the function is also very easy to obtain, which is in the form of $X^2=2py$. But, if these data points appear to be irregular, then it is difficult to find a single function. Modern people think to use a multifunction superposition method to simulate an overall function. The weight of each function is adjusted, so that the superposition function curve can pass through as many points as possible. Ptolemy recorded a large amount of data on celestial motion and then tried to simulate the function of the elliptical trajectory by superimposing multiple circular functions in order to include all the data he recorded. The fitting-function method is suitable for finding the law from a large number of discrete data records, which is the foundation of today's artificial intelligence and the basic mathematical method of machine learning. Many of the modern basic mathematical methods have existed in the past, but they were not put into use due to limited capacity.

Today, humans can restore history by digital mapping. Even in a game like My World, the computer can calculate the angle and length of each tile and reproduce the perfect three-dimensional image of the ancient walls that were built thousands of years ago. You will feel that all the splendid history of ancient Egypt, ancient Greece, and ancient China are reconnected with us. But, compared to a fading golden crown and silver belt, ancient people's wisdom of using data may be the most precious inheritance of mankind.

Data civilization is progressing, but most people are still unfamiliar about data. In everyday life, the concept of data is both familiar and unfamiliar to us. We get close to it, because everyone will learn some basic data and algorithms, such as addition, subtraction, multiplication, and division. After entering society, we inevitably deal with all kinds of documents, statements, and bills, no matter what job we take. But when we face various fashionable and complicated data, such as memory and high-tech resolution, we find understanding data, or even being aware of it, difficult. With the advent of big data, machine algorithms, and artificial intelligence, this abnormality has deepened.

So is data far away from our life? On the contrary, under the new technical conditions, the connection between data and our daily lives has never been so close. Our ancestors learned to store data in a structured way, but they were not as active and specific as we are today.

From calculators and cameras to home computers and smartphones, and to big data and artificial intelligence, we are constantly upgrading the way through which we collect and use data. Now, from the daily carbon emissions of a car to the monitoring of global temperature, from the analysis of everyone's online comment to the prediction of the voting trend in the presidential election, from predicting the rise and fall of a stock to observing and evaluating the development of the entire economic system, we can do everything. Data connects person with person, as well as people with the world, thus forming a dense network. Everyone is affecting the world, and everyone is being influenced by others. This dialectical relationship from micro to macro is like the phenomenon of quantum mechanics that occurs in all human beings, which answers countless questions. Traditional statistical methods have been unable to process such interaction data. So what can we do? The answer is to let the machine handle the data itself, and then we learn from the data. This is the essence of contemporary artificial intelligence.

As early as sixty years ago, artificial intelligence was studied by scientists seriously. Even though ordinary people are interested in artificial intelligence, there has rarely been any breakthrough during the rapid development of human science and technology after World War II. That is, until today, when we suddenly see that artificial intelligence has sprung up, breaking into our lives with new features, such as big data, AlphaGo, and Baidu unmanned vehicles.

If we compare the technology of artificial intelligence to a heart, then we can say that the heart has suffered from two congenital deficiencies: First, before the Internet, the amount of data that artificial intelligence could use was too small, which was "insufficient blood supply." Second, the lack of hardware led to inefficiency in computing power to solve complex problems, which is "insufficient mentality." Data is like blood; hardware is like blood vessels. These two problems were unsolved until the Internet advanced by leaps and bounds, the computing power of computers doubled in one year, and revolutionary changes took place in computing architecture. The rushing data blood enters every corner of the physical body: image recognition, speech recognition, natural language processing. Their eyes and mouths are open, ears are alert, and the heart of the machine is alive!

Data Writing Life Cycle

We have been deeply immersed in data. Computers, smartphones, and various smart appliances are collecting our words and actions; they know more and more about us through computational modeling, and the simplest daily activities such as watching news, engaging in sports and fitness, eating, listening to songs, and traveling have become the grand data festivals.

A smartphone can produce 1G of data for his or her owner in one day. This probably equals the total capacity of thirteen sets of *Twenty-Four History* (a multivolume history of China). Every day we use data to write our vast life cycle.

Unlike traditional data record definitions, this type of data is "alive." This kind of record is not an objective and absolute mathematical measure, nor is it a one-on-one historical writing. It is more like a natural extension of our body: intelligent agents listen to our speech, broaden our vision, and deepen our memories, even though it is forming another "me" in the form of data. If the smartphone has become a new organ of human beings, then the data is the "sixth sense" received by this new organ. The new brain that deals with this sixth sense is the rising artificial intelligence.

Big Data—Everything Is Data

Humans have been using data for a long time, and since the industrial revolution, data has become more commonplace. Then why has the concept of big data emerged only in recent years? Can big data do anything apart from recording and calculating more data? Making big data functional involves several characteristics.

Size

First, consider the bigness of big data. Undoubtedly, compared to the traditional data-storage methods, bigness is not bigger on the same order of magnitude, but bigger in a geometric dimension. Think about the 72 billion daily positioning requests on Baidu Map, and think about the number of clicks on the Internet, the number of words and images sent on social media every day. The amount of data collected by various big data platforms in one day can surpass the sum of words and images that humans have accumulated for thousands of years.

Multidimensionality

Second, multidimensionality is a factor. Multidimensionality means that big data can describe a thing in multiple directions and thus be more accurate.

In the movie *Jason Bourne*, a big data company helped the US Central Intelligence Agency (CIA) quickly track and locate suspects based on data collected from various dimensions, such as Internet data, traffic data, and historical archives. In reality, Palantir Technologies of the United States has helped the US government track down Osama bin Laden and provided counterterrorism information and warnings of social crises. The company is more often used to discover financial fraud.

Take the financial credit application as an example. When conducting traditional credit reporting, traditional financial institutions generally collect about twenty types of data, including age, income, education, occupation, property, and borrowing history. After they get a comprehensive score on the customer's repayment ability and willingness to repay, the credit limit is determined.

Internet companies adopt big-data methods, and the dimensions they acquire can actually startle traditional banks. BAT (Baidu, Alibaba, and Tencent) has opened its own financial services. Due to the comprehensive and huge user data, the group can query customers' various online records, such as abnormal behaviors like bulk application for loans; it can also compare customer information with Internet global information. The records are scrutinized for fraud patterns; further, the customer's consumption behaviors and habits can be analyzed, and then the company will know the applicant's repayment ability with the reported income. Of course, respecting the privacy of the user, the data will not be disclosed. The convenience for users is that the wait time for the credit investigation is greatly shortened. Big data can retrieve and review the original information of more than ten thousand applicants in a few seconds and quickly check tens of thousands of indicator dimensions.

Examining the creditworthiness of a stranger has been similar to the story of a blind man feeling an elephant, touching only some part of it, and concluding what the elephant is like. The traditional method is based on the evaluation of a customer's credit "elephant" through twenty "blind people," whose understanding may be flawed. The multidimensionality of big data is like tens of thousands of people simultaneously "touching the elephant" and then putting all their feedback together. The more dimensions there are, the more accurate the conclusion is.

How Data Is Stored

Third is the ability to process unstructured data. The most basic numbers, symbols, etc., in structured data can be stored in the database with fixed fields, lengths, and logical structures and presented to humans in the form of data tables (think of the common Excel tables), which are very convenient to handle. However, the Internet era has produced a large amount of unstructured data. The data of pictures, videos, audio, and other content is huge in volume but has no clear structure. For instance, we can only regard image data as myriad pixels on a two-dimensional matrix. Unstructured data is growing rapidly, presumably to account for 90 percent of total data in the next ten years. Big-data technology can calculate and analyze a large amount of unstructured data through image recognition, speech recognition, natural-language analysis, and other technologies, thus greatly increasing the data dimension.

The amount of unstructured data is far greater than structured data; unstructured data takes up huge resources and has broad application prospects. For example, in the past, personal identification in public places such as airports could only be verified based on identity information provided by passengers. After the application of face recognition, speech recognition, and other technological measures, big data can directly check the passenger through the camera, increase the dimension of personal identity judgment, and conduct accurate and efficient security checks.

Duration

Fourth, big data is a continuous flow, like time. It never returns to the past, just as people could not step into the same river twice. This is because the amount of data is too large to be fully stored. On the other hand, big data is related to the actions of humans, which are constantly changing. Therefore, Baidu's Big Data Lab proposed a concept called "time and space big data."

The map is the mother of time and space big data. Baidu Map has a road congestion warning function. If the road is clear, it will be displayed in green; if it is congested, it will be displayed in red to notify the user to choose another route. This is a concise example of how we interact with data.

If we have two routes A and B, and route A is congested, and route B is smooth, then we will choose route B; when more and more drivers choose route B, it will become congested, and route A will be smooth again. Everything keeps changing. Relying on the positioning function of smartphones, Baidu Map can change the current road condition monitoring result in real time and accurately report the current road condition to each user. Data-visualization techniques and various assessment methods can be used to describe the daily pulse of a city, such as changes in the flow of commute, as if the city's breathing. Apart from recorded data, data is only valid at the moment. It is impossible to store all the data; even the land of the entire city may not be enough to pile all the hard disks. It can only be applied immediately, and it disappears after being used.

Keeping up with time data is really challenging. In November 2016, Baidu officially accessed the children's missing-information emergency-release platform of the Ministry of Public Security. Whenever a child is missing, Baidu Map and the mobile phone Baidu app will accurately forward the important information, such as the name, appearance, and time of missing child, to users in the vicinity, so that they can participate in the search process without delay. After the child is found, Baidu Map and mobile phone Baidu app will also update the closing time to let people keep abreast of progress. The less time it takes to provide the information to the users, the more hope the anxious family will get.

Repetition

Last but not least, the bigness of big data is endless repetition. For speech recognition, people's repetition of same statements helps the machine to fully grasp human speech by repeatedly identifying the nuances of human speech. Also, people's recurring movements can help the system to capture the laws of urban movement. The mathematical meaning of *repetition* is "exhaustion." In the past, human beings could not grasp the law of a thing through exhaustive methods. They could only use sampling to estimate or use simple and clear functions by observation to represent the law of things, but big data made the "stupid method" of exhaustion possible.

Quantitative change promotes qualitative change. In the field of machine intelligence, the amount of data and the speed of processing can directly determine the level of intelligence. Google's story of improving the quality of translation through the amount of data is no longer a secret.

In 2005, the National Institute of Standards and Technology held the annual evaluation of machine-translation software. Many university institutions and large companies have applied for research funding for machine translation from the US government, and they needed to take this evaluation. Teams or companies that did not receive government funding have also joined, and Google was one of them. Other participants included IBM, the German Aachen Institute of Technology, and many famous companies in the machine-translation industry, all with strong experience in machine translation for many years, but Google was the first.

However, the results of the evaluation stunned people: Google won first place and scored much higher than other teams. In terms of Chinese-English translation, Google's performance rated at 51.37 percent of the BLEU score, and the second- and third-place companies only scored 34.03 percent and 22.57 percent. Finally, Google announced its secret: use more data! Google didn't just use twice as much as other teams, but more than ten thousand times more! Google can collect massive amounts of human bilingual data on the Internet through search engines. For the same sentence, there will be many different Chinese translations, and the computer will use this repetition to find the most commonly used translation. Without any other changes, Google has trained and transformed a product ahead of the other machine-translation methods over a generation, solely relying on increased data samples. In fact, the secret of Google's success is its super exhaustion ability.

The data advantages of Internet companies such as Google and Baidu are wide-ranging. In addition to translation, these advantages can be easily copied to other fields, such as speech recognition and image recognition. As a small game, Baidu's app for writing poems combines big data and artificial intelligence. According to Zhongjun He, Baidu's chief architect and head of machine-translation technology, traditional poetry-writing software generally uses statistical models to generate the first verse by a given keyword and then generates the second verse, repeating the process until the whole poem is completed. Baidu writes poetry in another way: the user inputs any words or sentences, and the system combines the big data in Baidu's search engine to deeply analyze and associate the user expression and derive the related keywords with high relevance. For instance, the user enters the word "West Lake" at random, and Baidu's poetry system analyzes the data of a large number of poetry and prose to find out which keywords should be included in a poem describing "West Lake." Here the keywords may include "broken bridge and unmelted snow," "misty rain," "drooping willow," and so on. Next, a poem is generated by deep neural-network technology, based on each subject word. These keywords are equivalent to the outlines we often use in writing. Creation according to the outline can ensure that the whole poem is unified in the artistic conception and the content of the verses is logically smooth. In the past, people said that single sentences of the poems written by the machine seemed to be good but the overall artistic conception was far from satisfactory, and now that problem can be effectively remedied. Each sentence is generated by machine-translation techniques. The first sentence of poetry is "translated" to generate the second sentence, then the second sentence is "translated" to generate the third sentence, and so on. We use "West Lake" as input, and the seven-verse poem generated by Writing Poems for You seems beautiful and logical.

Human Data Mirroring

Humans are becoming increasingly picky, spoiled by technology products, and big data can provide a dazzling color relative to boring choices. In the past, TV sets did not respond to our emotions, but now video websites are patiently and carefully collecting every kind of feedback from us: downloads, closing or fast forward; all kinds of actions are recorded. Then big data calculates various indicators, such as our preferences and spending power.

The American TV series *House of Cards* was popular for a time. The politicians on TV are playing cards, but behind them is big data playing invisible chess. It is produced by the famous American Internet TV company Netflix, which is familiar with the power of big-data analysis. In addition to the foregoing user behavior described, it also tried to collect viewing time, viewing equipment, viewer number, and viewing scenes, analyzing the starring actors and directors of users' favorite programs. Through big-data analysis, it was concluded that a show like *House of Cards* would be hot, so Netflix purchased the remake copyright from BBC (British Broadcasting Corporation) at a high price, and predicted Kevin Spacey to be the most suitable starring candidate. The result proves that Netflix's bet on *House of Cards* was completely correct. When we sighed in front of the screen that the president played by Spacey had the wisdom to control everything, we did not realize the power of the "data president."

The current US president, Trump, made full use of the data for the election. According to media reports by Bloomberg and other media, his technical team used public data on Facebook, Twitter, and other platforms, such as giving a thumb up, forwarding, and collecting behaviors, to accurately describe the portraits of voters and push individualized campaign ads to them. Each Twitter and Facebook message is sent targeted with different content directed to different netizens.

Baidu Brain is also good at developing accurate portraits of the users by big data. In 2016, the producer of the popular movie *Warcraft*, Legendary Pictures Productions, cooperated with Baidu Brain and accurately pushed the film ads to potential audiences according to the massive analysis of users. Although the film had a bad box-office draw in the North American market, it sold $221 million in China. When the *Warcraft* fans shouted, "For the Horde!" in the theater, perhaps it was big data that quietly gave them the Force.

Chinese people always say, "Food is god for the people." Regarding movies, how to eat well is a hot topic for the entire people. In 2013, Baidu published "Ranking of China's Top Ten Foodie Provinces and Cities," which became very popular on the Internet. This list uses the big data of Baidu Knows and Baidu Search, according to netizens' 77 million Q&As about eating. It summed up the different eating habits and characteristics in various places.

A lot of interesting phenomena have been unearthed in the massive data. "The fruit to help lose weight most quickly" has been asked by as many as three hundred thousand people. It seems that many netizens are still thinking about their bodies while eating. "The crab was still alive last night but died today; can I still eat it?" There are as many as sixty thousand responses to this question. It can be seen that Chinese foodies have a particularly high passion for crabs. Of course, there are more daily questions such as "Can I eat *X*?'" and "How to cook *X*?" Just the simple question "Can spinach and tofu be eaten together?" has triggered countless discussions.

These problems alluded to by the questions are huge and seemingly confusing. But repetition is exactly the beauty of big data. Big data can capture deeper meaning. For example, netizens in Fujian and Guangdong often ask questions about whether certain insects are safe to eat, while netizens from the Northwest are quite uncertain about how to cook seafood. Different users care about different ingredients and practices. Baidu Big Data summarized the attribute of foodies from each province and city. Behind this, big data considers the geographical location of netizens, the time of questioning and answering, the information provided about eating or cooking, and even the various dimensions of mobile-phone brands used by netizens.

In addition to portraying human attention to information, big data is even constructing our bodies. Nowadays, many people are familiar with wristbands, which analyze our health condition and make recommendations based on daily exercise data, such as walking steps, calorie consumption, sleep duration, etc. In the future, we can upload our personal data and use big data to detect the possibility or potential threat of various diseases in order to better prevent them.

There are many examples of big data in life. Most of the advanced Internet products we use today, whether computers or smartphones, are more or less related to big data. When we use these services without thinking, we have already invited big data into our lives. It silently looks at every detail of our lives, subtly encouraging and advising us to make choices and strengthen our role.

Breakthrough: Machine Learning and Artificial Intelligence

In 1950, Alan Turing created a test method for machines—the later famous Turing test. The legendary scientist believed that if a machine could talk to humans (via telex equipment) and could not be identified as a machine, then the machine could be considered intelligent. This simplification convinced people that thinking machines are possible, and the Turing test has been an important criterion for artificial intelligence until now.

This standard has diverged: as long as the machine behaves like a human, we don't have to worry too much about the machine's operating rules. Some people have proposed ways to let the machine learn the rules by themselves so humans don't have to worry about them.

In 1949, Donald Hebb took the first step in machine learning based on the learning mechanism of neuropsychology, creating a method that was later called Hebbian learning rule. Hebb believed that the learning process of neural networks occurs at the synaptic sites between neurons. The intensity of synaptic connections changes with the activity of neurons in front of and behind the synapse. Correct feedback will strengthen the connection between the two neurons. This principle mechanism is similar to Pavlov's conditioning experiment with dogs: each time before feeding, the experimenter rings the bell; after some time, the dog's nervous system will connect the ringing with the food. Hebb used a set of weighting formulas to stimulate the human neural network, with the weights representing the strength of the connections between neurons. He created a set of methods for the machine to easily distinguish between things. For each piece of data, let the decision-tree program make a judgment, reward it if it is right (improve the weight of the function), and punish it if it is wrong (reduce the weight of the function). He used this method to create a classifier that can extract the statistical properties of the data set and classify the input information into several classes according to their similarity. It seems like how human beings observe and summarize and distinguish things when observing a certain phenomenon, but this "observation" of the machine is more like a conditional reflex achieved through training. It is not based on internal thinking like humans do but pays attention to the correlation relationship contained in the data, instead of the causal relationship in human thinking.

In the next decade, research on artificial intelligence became increasingly intense. In 1952, IBM scientist Arthur Samuel successfully developed a checker program that could improve itself. He coined the term *machine learning* and defined it as "a field of research that provides computing power without explicit programming."

In 1957, Frank Rosenblatt proposed the perceptron algorithm, which became the basis for the development of neural networks and support vector machines (SVMs) in the future. Perceptron is a kind of algorithmic classifier, a linear classification model. The principle is to separate the data by continuously training trial and error to find a suitable hyperplane. (The hyperplane can be understood in this way: in the three-dimensional coordinate space, the two-dimensional shape is called a plane and can divide the three-dimensional space. If the data is multidimensional, then in the N-dimensional coordinate space, the N-1 dimension is a hyperplane, which can divide the

N-dimensional space.) As you enter the two piles of balls labeled CORRECT and WRONG, the perceptron can find the dividing line between the two piles of different balls for you.

A perceptron is like a neural network with only one layer between input and output. When faced with a complicated situation, it is powerless. For example, when the "correct" and "wrong" balls are mixed with each other, or when a third kind of ball appears, the perceptron cannot find the boundary of the classification. This makes it difficult for the perceptron to make a breakthrough on some seemingly simple issues.

Nowadays, humans don't have input rules (programming) but let the machine look for rules by itself, so that the machine can use its own intelligence. Today's artificial intelligence is developed on the basis of machine learning, only its speed of growth being limited by hardware and methods.

If multiple computers and multiple chips are set up in a network for machine learning and have multiple chip network layers, then they will be categorized as so-called deep learning. In the late 1970s, Professor Geoffrey Hinton and others had discovered that if a multilayer neural network can be implemented, the pattern in pattern can be found step by step, allowing the computer to solve complex problems by itself. At that time, they developed a "backpropagation" algorithm neural network. However, the complexity of multilayer neural networks led to a significant increase in the difficulty of training, and the lack of data and hardware-computing capabilities became a constraint.

From the mid-1960s to the end of the 1970s, the pace of machine learning was almost stagnant. This situation did not improve until the 1980s. With the rapid development of computer performance and the advent of the Internet, artificial-intelligence research finally became more powerful. In the 1990s, modern machine learning was initially formed.

The Internet was put into commercial use in the 1990s, which led to the development of distributed computing methods. Supercomputers are expensive, and distributed-computing technology takes advantage of the large quantity, allowing multiple ordinary computers to work together, each computer undertaking part of the computing task and aggregating the whole results, which can outpace the supercomputer. The distributed structure is suited to the increasing amount of data.

Computer Neural-Network Growth and Deep Learning

As traditional artificial intelligence relies blindly on the rule of model input by scientists, it can only work effectively when solving problems with relatively clear rules. For example, Deep Blue, which defeated world chess champion Kasparov, is such an example of artificial intelligence. However, when faced with the simple problem of recognizing a picture, which humans can learn in the infant stage, such artificial intelligence would be in a quandary, because this kind of cognitive problem only has a vague concept, with no clear and simple rule available. The computer neural network does not require humans to declare the rules in advance—it can identify the patterns (rules) from a large amount of basic data by itself.

As the name implies, the neural network resembles the human brain and consists of many neurons. Each neuron is connected to several other neurons to form a net. A single neuron only solves the simplest problems, but, when combined into layers, neurons can solve complex problems.

Geoffrey Hinton believes that the traditional machine-learning method used only one layer of chip network, so its processing efficiency became very low when dealing with truly complex problems. The core idea of deep learning is to increase efficiency by adding the number of neural-network layers and to abstract and simplify the complex input data layer by layer; in other words, the machine solves the complex problem by dividing it into subsegments. Each layer of neural network solves its own problem, and the result of this layer is passed to the next layer for further processing.

With one layer of neural network, simple patterns can be found; with multiple layers of neural networks, patterns in patterns can be found. Take face recognition as an example. The first layer of the neural network only focuses on the image areas with sides of dozens of pixels, from which some shapes (shapes are patterns)—eyes, noses, and mouths—are recognized. Then these already-recognized shapes are handed over to the next layer of neural network, which finds a bigger pattern from the existing recognition results and combines them into adult faces. Stated more mathematically, the current popular deep neural networks can be divided into CNN (convolutional neural network) that responds to data with spatial distribution and RNN (recurrent neural network, also known as circular neural network) that responds to data with temporal distribution.

CNN is often used for image recognition. The first layer of the network is trained to accomplish a small target for recognizing local independent modules of the image, such as a square, a triangle, or an eye. At this level, humans input a large amount of image data, to allow the layer to only discern the basic edge of the local image, i.e., nothing besides a pixel. Each of the following layers looks for a higher-level pattern from the information derived from the previous layer. This method simulates the way that human eyes compose information, which discard minor details and prioritize certain salient patterns. For instance, several small pieces and a circle combine to form a face; no matter where it appears in the image, the human eye will first pay attention to the face, instead of all the single parts of the image.

RNN is often used for speech recognition and natural-language processing because speech and language are data distributed according to time—the meaning of the next sentence is related to the previous one. The RNN network can remember historical information. Suppose we need to develop a language model and use a preceding sentence to predict following words. Given that "I was born in China in 1976. My college major is mathematics. I speak fluent _____" the last word of this sentence is obviously Chinese (the language that Chinese people speak), which is very simple to grasp for humans. But computer neural networks need to get the previous information of "China" to finish the work, which requires a cycle of design, so that the neural network can have a temporal depth.

Deep neural networks have greatly optimized the speed of machine learning and made breakthroughs in artificial intelligence. On this basis, great progress has been made in image recognition, speech recognition, and machine translation. Speech sound input is much faster than typing; machine translation basically allows us to understand a piece of foreign-language information; image recognition can precisely find one person from a pile of adult photos by means of his or her photo as a teenager, and it can even restore very vague photos to very clear and accurate ones.

Artificial intelligence based on deep learning is different from the previous artificial-intelligence principle but shares similar logic with what we know about data mining: get results first and look back in reverse for patterns. This process is called training.

With simple mathematics knowledge, we can explain the basic thinking modes of machine learning, training, and deep learning. This method is comparable to the Copernican reversal in the field of mathematics. A simple function is used as an example to illustrate this reversal.

In the past, we solved mathematical problems by generally knowing formulas (functions), then using input data to find results. Take the function y=ax+b as an example. If you know y=2x+1 and let x=1, you can find y=3. Here x is input, and the resulting y is output.

A higher-order mathematical ability is to know the formula and the output and use them to find the input value. For example, if y=2x+1, let y=5, and find x.

One step further, we will touch machine learning. When we don't know the coefficients a and b, but know the values of y and x, we need to find a and b; that is, knowing input and output, we can find the function coefficients. In the y=ax+b function, we only need to know the two sets of x and y values to confirm a and b.

Further, suppose we have a set of input and output data, but we don't know the form of the function at all. What should we do? This requires a constructor. For example, it is known that x=2, y=5, and find f(x). This cannot be calculated with very few input and output data; f(x) may be 2x+1, 1x+3, or even x2+1, and countless other cases. However, if the number of x and y are sufficient, the mathematician can adjust the weight of the formula through the approximation calculation method and approximate the function.

The problem is that with extremely large and complex data, which is generated in our modern production and life, we need to be rather highly efficient if we want to obtain the functions contained in the data. The human brain is no longer competent, but it can be handed over to the computer. The fitting function shows its magic here. The deep-learning neural network simulates the neural nodes of the human brain. Each node is actually a function regulator, and numerous functions are connected to each other. Through various mathematical methods such as matrix, optimization, and regular expressions, the deep-learning process continuously adjusts the weight of each function coefficient. When the data is adequate and the construction principle is appropriate, the evolving function will be more accurate in fitting most of the data, then we can use this set of functions to predict what hasn't happened yet. This process is what we call "training."

When Andrew Ng was working at Google, he led his team to successfully train the famous computer cat-identifying system. If we use the old-fashioned symbolic artificial-intelligence method to program, then first we need to carefully define the cat, such as sharp ears, round eyes, straight beard, four legs, long tail, etc.; convert these characteristic definitions into proper functions; input these functions into computers; and then present a picture to the computer. The computer breaks down different elements of the picture and compares these elements with the rules defined in the program. If it conforms with the characteristics of sharp ears, round eyes, straight beard, four legs, long tail, etc., then the picture is of a cat.

The method of machine learning is quite different. Scientists do not write the definition of a cat in advance, but let the computer find it. Scientists just "feed" a large number of pictures to the computer and let the computer output tags, either cats or not cats. Numerous pathways are generated in the neural networks that can identify cats. Just like human brains, each pathway outputs its own result. If it is correct, scientists will increase the weight of this pathway (can be interpreted as a green light); if it is wrong, the weight will be reduced (can be interpreted as a red light). After enough attempts, such as testing one hundred thousand pictures of various cats, the weighted neural pathways form a recognition device (a complex set of function linkages). Then the cat in the new picture can also be identified without scientists telling the result of the identification. The more training data there is, the more complex but accurate the set of functions becomes.

This is "supervised learning"—relying on a large amount of tagged data. The cat-identifying project led by Andrew Ng can even learn from the scratch, and cats can be identified without labels. When researchers showed millions of static cat pictures to the neural network, the neural network obtained a stable model by itself. From then on, it identified the cat's face without any hesitation, like all children.

Andrew Ng's doctoral student Quoc Viet Le has written a paper based on this, which shows that machine learning can also identify the original unlabeled data and establish his own knowledge model. Its significance is by no means limited to the identification of cats.

More than two decades ago, intrigued by the beehive effect, Kevin Kelly narrated his opinion in his outstanding scientific book *Out of Control*. He used this method to predict the emergence of new technologies such as distributed computing, even when he may not have seen the machine-learning principle behind the beehive effect. Each bee's movement is random, but the bees in the hive can always fly in one direction. A large number of bee's respective actions (inputs) are aggregated into a total movement (output), and the logic (function) behind that is the beehive effect. The information movement in the computer neural network is like the supersonic flying bee colony collecting data pollen. In their seemingly frantic trajectory, a cat's face is highlighted. Baidu's ability to identify cats has gone far beyond human's. It can even accurately distinguish different species of cats.

So, for humans, machine learning often forms a black box. Some people have warned that this kind of black box that transcends human understanding will bring danger because we don't know how the machine thinks and whether it creates dangerous thinking. But, more often than not, deep learning can bring unexpected surprises.

Anecdotes about Baidu

An engineer at the Baidu Speech Recognition Development Team once related an interesting story: When a speech sound team member tested the speech-recognition program at home, he inadvertently sang a few lyrics, and the lyrics were accurately identified. This surprised him, because other companies' speech-recognition technology can't do that. The Baidu team members did not train for this form of unaccompanied singing, nor did they set that goal. They don't know how the system does it. The training data must have reached a sufficient level. The program has cultivated this amazing skill in the process of continuous training and learning.

People tend to know about changes in the world slowly and feel behind the times. In the days without deep learning, the world seemed to be all right. But some of the invisible burdens are being silently undertaken by some people. Zhou Kehua, a serial killer, had come and gone like a shadow for more than a decade. In order to seize him, the public security department mobilized almost all video surveillance materials to discover his traces. At that time, how did the police officers retrieve videos? By naked eye! They needed to watch the video clips hundreds or even thousands of hours long; some police officers even fainted at work. However, visual recognition based on deep-learning techniques changes that. At present, advanced monitoring systems have strong artificial-intelligence support. After the training with big data, they can instantly recognize faces, license plates, models, etc., from the video and semantically facilitate human retrieval. So, just give the computer a few photos of the suspect, and the neural network can quickly find the suspect-related footage from the massive video for human reference. Security enterprise Yushi Technology has developed such a smart camera system; when combined with Baidu Map, it can quickly locate the path of suspects or vehicles.

Deep learning has already changed our lives in many invisible aspects. In order to collect and maintain map information, it is necessary to capture images along the road through collection vehicles. Traditionally, a collecting car needs two staff members. The collecting process is divided into two parts: internal work and external work. The external work is to drive the vehicle and record the things along the way. In addition to video recording, the copilot is responsible for recording with speech. Every time they pass a certain place, one person has to say that there is a probe in front, a traffic light, or four lanes, or the person needs to tell the driver to turn left, go straight, turn right, and so forth. Under this traditional way, the staff members need to record all the things they see in the form of video and sound and then send the data to the data-processing center. The center's internal personnel record and compare the data by the minute and finally mark the elements of the road on the map, which is basically labor-intensive work.

After applying the intelligent image-recognition technology, we first use the deep-learning training machine to identify the road elements, such as traffic lights, lanes, probes, etc. After that, we only need to directly submit the panoramic image taken along the road to the machine for identification, and then we get the complete map information. This greatly saves manpower and greatly improves efficiency and accuracy.

In addition to software algorithms, there is a story about hardware on deep learning. There have been many inventions in history that deviated from their original intentions in later applications. For example, although an explosive, nitroglycerin can be used for first aid in heart disease. The original attempt to invent artificial synthetic substitutes for strategic materials ended up in the invention of plasticine. In the field of deep learning, the role of GPU has also been changed. The GPU was originally a graphics card that was used to render images and speed up graphics calculations, but later it became the main hardware for deep learning. Because the graphics chip has more floating-point computing power than the CPU and was originally used to process matrix data such as images, it is very suitable for the calculation of data in the machine-learning field. In the early days, when Andrew Ng's team took the lead in using the GPU for machine learning, many people did not understand. But today it has become mainstream.

But the most impressive anecdote is still about a search engine.

Search Engine: AI's Thin Destiny Curve

Baidu's focus on artificial intelligence caused some incomprehension in the past. Why does Baidu show a special preference toward artificial intelligence instead of countless popular fields from PC to mobile-Internet applications such as e-commerce, games, social media, and communication?

The answer to the question may be contrary to what many people believe. Instead of saying that Baidu chose artificial intelligence, we would rather say that artificial intelligence chose Baidu. This is the mission in Baidu's genes. Failure to live up to this mission will be the loss of Baidu, China, and even the world.

Everything Comes from Searching

For general users, search engines are just a tool to help them to find the information they need; for websites that provide the content, search engines are mediums that help them to deliver their content to users in need. In this process, the search engine first "listens" to the user's needs; that is, it determines what the keywords typed in the small search box want to find. Next, search engines "retrieve" a large amount of content and pick out the results that best meet the requirements.

Let's examine this process. Is it very similar to the deep-learning model? Inputs and outputs are here, and even every search can be seen as a training exercise for search engines. So, who tells the search engine that the results shown are good or bad? The user does. The user's click is an answer. If the user does not click on the top results, but clicks on the result of the second page, this is a demotion of the recommendation of the system.

In this process, the search engine not only improves the accuracy of the recommendation but also knows more about hits and misses of the included web pages and gradually learns to distinguish web pages like humans. Initially, it could only read page elements such as titles, keywords, descriptions, etc. But, now, search engines like Baidu can identify which is hidden false information, which are advertisements, and which is truly valuable content.

The action of people getting information through search engines is the process of dialogue between people and machines. Unlike previous human-computer interactions, this process is based on natural language. Compared with image recognition and speech recognition, natural language processing (NLP) is the core technology of search engines.

Wang Haifeng believes that the ability to think and acquire knowledge has made today's human beings. This ability needs to find objects and methods of thinking through language and is externalized as our ability to see, listen, speak, and act. Compared to all these abilities, language is one of the most important characteristics that human beings use to distinguish themselves from other creatures. Visual, auditory, and behavioral abilities are not limited to human beings but belong to all animals. Many animals even have better visual, auditory, and behavioral abilities than humans, but language is unique to humans. Summarization, refinement, inheritance, and thinking of knowledge based on language are also unique to human beings.

From the beginning of human history, knowledge has been recorded and transmitted in the form of language, and the tools used to write language are constantly improving: from oracle bone to paper and then to today's Internet. So, both Baidu and Google believe that natural-language processing is a very big challenge for the future of artificial intelligence. In contrast, speech recognition, such as speech to text, or text to speech, is actually a process of simple signal conversion, but language is not like this—it involves human knowledge and overall thinking rather than simple conversion.

Projects like AlphaGo astonish ordinary people, and we think it is a big achievement. But actually, we can't ignore its characteristics that are based on complete information, clear rules, and closed and specific space. The intelligent system trained to play Go is not good at playing chess. In comparison, the processing of natural language is a more difficult problem to solve. For playing Go, there is almost no uncertainty as long as the computing power and data are sufficient, but there are too many uncertainties in language problems, such as the diversity of semantics. For computers to "understand" and generate human language, scientists have already done a lot of work. Based on the accumulation of big data, machine learning, and linguistics, Baidu has developed a knowledge map; built a system for question and answer, machine translation, and dialogue; and acquired the ability to analyze and understand queries and emotions.

The knowledge map can be divided into three categories based on different application requirements: entity graph, attention graph, and intent graph.

In the entity graph, each node is an entity, and each entity has many attributes. The connection between the nodes is the relationship between the entities. At present, Baidu's entity graph already contains hundreds of millions of entities, tens of billions of attributes, and billions of relationships, which all derive from a large number of structured and unstructured data.

Now let's look at an example: someone searches for the ex-husband of Dou Jingtong's father's ex-wife. (Dou Jingtong, a.k.a. Leah Dou, is a Chinese singer-songwriter.)

The relationship of the characters contained in this request is very complicated. However, our reasoning system can easily analyze the relationship among entities and finally get the correct answer.

Baidu's natural-language processing technology can also analyze complex grammar and even identify ambiguities in sentences, not just literal matches.

Let's look at an example: Who is Liang Sicheng's son? Whose son is Liang Sicheng? (Liang Sicheng was a Chinese architect.)

If we use traditional keyword-based search techniques, then the two queries will get almost the same results. However, through the analysis of semantic understanding techniques, the machine can find that the semantics of these two sentences are completely different, and correspondingly, completely different answers can be retrieved from the knowledge map.

Consider a third query: Who are Liang Sicheng's parents? Literally, this is different from the second query, but through semantic understanding, the machine can find that the two sentences are looking for the same object.

Deep-learning technology further enhances natural-language processing capabilities. Baidu has applied the deep neural network (DNN) model to search engines since 2013, and, so far, this model has been upgraded dozens of times. The DNN semantic feature is a very important feature in Baidu search. In fact, not only have the relevance of search results become higher, but also the chapter understanding, focus perception, and machine translation have been greatly improved.

The technical foundation required for searching is also required for artificial intelligence. For example, in terms of cloud computing, Zhang Ya-Qin, former president of Baidu, believes that searching is the largest cloud-computing application. Without the cloud, there would be no way to search. Baidu was born in the cloud.

Search Engine Continues to Evolve

With the rise of mobile Internet and artificial intelligence, the form of searching has changed a lot. For example, the search portal has changed. In addition to the search through the web search box, searches based on different platforms and hardware are also increasing, and speech sound or image search replaces part of the text search. While people are actively searching for information, information is also recommended to those who need it. Many people judge things from their appearance and think that this process is a challenge to search engines. But Wang Haifeng believes that developers of search engines have been simultaneously aware of this changing process.

Many Internet companies agree that "information looks for people." However, people looking for information and information looking for people, or search and feed, are not one or the other, but complementary. They play different roles in different scenarios at different times, but they perform their own duties and cooperate with each other. For example, when you need to look for some information, sometimes your friend makes a recommendation, and sometimes the system guesses your preferences and makes a recommendation. Suppose someone recommends an article to you. When you find a word that you don't understand well while reading it, you need to search to find the meaning of that particular word. Of course, the machine will also guess which words users may be interested in. When a piece of content is not so popular, you have to look for it in the search engine. In different scenarios, the user's needs for search and feed will be converted to each other, and how to judge these scenarios is a test for the system's intelligence. The more data and technology reserves there are, the better it may perform.

With technical reserves and data, it is technically not difficult to make a feed. But it is more difficult to start from feed and make up for the lack of search and data. The Baidu search engine collects and analyzes hundreds of billions of web pages. Such large-scale data is the necessary guarantee for Baidu to continuously improve the performance of its feed products.

Search engines continue to evolve with regard to data torrents, and feed is just the next step. Eventually, a ubiquitous search engine plus a recommendation will form. More and more intelligent machines can draw inferences about other cases from one instance. In the end, users will only say a few words, and the machine will know the whole meaning that the user wants to express. In addition, machines can automatically analyze the user's location, identity, habits, etc., and by using this information machines can easily determine which search results should be provided to users. In the future, in many cases, we won't have to start a search. A search-engine-based feed will guess and push the information we need. Imagine, for example, when having a meal at a restaurant, the search engine has inferred the user's next arrangement based on the user's previous search content. Even if the user has not asked, the search engine will voluntarily collect the information that the user might need later, such as what movies are currently on screen, where is the nearest movie theater, etc. Baidu has tried this idea in some of its products. Information of less interest does not appear in the feed, but will be stored reasonably, like an invisible library for users to explore later. Intelligent search engines are growing with us.

Searching Is the Largest Artificial-Intelligence Project

Search engines work without stop, like a mirror image of a human's learning spirit, collecting and processing large amounts of data at all times, grabbing pages and content on the entire Internet, visiting everything, including e-commerce, social media, and news portals.

The search engine is a seeder, a laboratory, and a digital collider. Combining speech recognition, image recognition, and machine translation, it can collect more valuable data through the actual use of a large number of users, which in turn helps the neural network to optimize the training effect and to form a benign development loop.

The development of natural-language processing technology will bring more surprises in the future. Writing financial and sports news within a certain format will be possible; even in literature, the poetry of the Tang Dynasty written by the machine will be like the real poems. When watching basketball and football games, the commentary robot will not only quickly report the game but also answer many people's questions at the same time. This is a bit like the smart program Samantha in the sci-fi movie *Her*, who falls in love with countless people at the same time. Love is probably the deepest language, thought, and emotional communication of human beings. Samantha is a high-level symbol of natural-language processing technology, depicting the deep relationship between humans and machines. Perhaps in the future, search engines will be like Samantha and exhaust symbolic information to break into the gap between language and meaning, which is beyond human imagination.

Strictly speaking, artificial intelligence is a kind of physical work, and it must have enough physical strength to withstand such huge data and calculations. In general, for colleges and universities or smaller Internet companies, the threshold of data volume and hardware cost limited the development of artificial intelligence a lot. Even if we exclude the purchase cost of hardware such as the CPU and GPU, the cost of running the hardware is high. AlphaGo costs $3,000 for electricity in one round of a Go game. In addition to traditional servers, bandwidth, and other infrastructure, Baidu now has hundreds of GPU servers that support artificial intelligence. Up to sixteen GPU cards can be installed on the highest-configured servers. On the basis of all this, the data reserve, hardware foundation, market scale, and talent team are coordinated to maximize the advantages. What we pursue is not the one-time gains, but the largest and most basic artificial intelligence platform, which works hard to help human beings achieve the "know more, do more, be more" initiative.

Artificial Intelligence Is the Fate of Baidu

It can be said that artificial intelligence is an intrinsic appeal for the companies like Baidu and Google, and it is also the appeal of the Internet, mobile Internet, and data explosion itself. It is very difficult for other domestic companies in this field to compete with companies like Google and Microsoft with such large-scale advantages. It is the unshakable responsibility of Baidu to establish an infrastructure and talent base.

To pass the artificial-intelligence torch to more people, create real value, make life better, and make the nation stronger gives Baidu workers motivation and a reason why Baidu can assemble many artificial-intelligence scientists.

Lin Yuanqing originally studied artificial intelligence in the NEC American laboratory. The outstanding conditions, atmosphere, and academic power of NEC helped him focus on research and publication. But he finally left the familiar environment and chose Baidu. He says the most important reason is that as an artificial-intelligence researcher he feels it is a very important part to practically apply the deep-learning technology. At present, there are more than 700 million Internet users in China, and more than 1.2 billion mobile phone users, which tops the world. How can we let users enjoy the changes brought by artificial intelligence and participate in it? The value of this exploration can affect the lives of all people in China. He eagerly felt that "this is the best moment, the most promising opportunity for artificial intelligence. It is a pity to miss it."

But artificial intelligence never rests. When human beings are sleeping, they are still rushing in the machine world, and in the endless cycle, they will succeed and fly to the world!

Here I want to end this chapter with a passage from a famous philosophical professor, written in the 1990s:

> In heaven, humans are not human. More precisely, humans have not been placed on the road of human beings. Now, I have been thrown out for a long time, flying through the void of time in a straight line. In a deep place, there is still a thin rope that binds me, and the other end is connected to the paradise where the clouds are covered in the distance. The individual soul is not her own choice, but the thin line from paradise is tied to her, controlling her body. It is impossible for Violica to find a passion for life. She can only find her life enthusiasm from herself, which means to discover the thin line that binds her body to the shadow. The thin line from the paradise determines the life direction of Violica's body and the burden of the individual soul, and makes her feel the individual destiny. The so-called individual destiny is nothing more than a person feeling that only such a passion for life can let her have a feeling of a beautiful life, and then she has the happiness of her own life, so that she must live like this.[4]

---

4   Liu Xiaofeng. 1999. *The Unbearable Body.* Beijing: Huaxia Publishing House.

4

CHINA BRAIN PROJECT: BOTTOM-UP SUPER ENGINEERING

The Human-Machine World Needs a New Brain Urgently

Many artificial-intelligence scientists have a professional background in both biology and computer science, which is probably a microcosm for the development of intelligent biology.

The earth is like a biological computer. The evolution of life is the long process of continuous iteration of various biological "programs." Due to the effect of the natural environment, inorganic substances gradually aggregate into organic molecules, which evolve and combine to form protein molecules. Protein molecules carrying life information are like single data bytes, and the random combination of large quantities can exhaust a variety of forms. Some of them can not only swallow and spit substances but also metabolize and copy themselves—life is thus created.

The most basic units of life carry their own genetic codes, the "programming language" of life. Nature used this language for various creations, genetic-code mutations and combinations, that generate various new structures, and it evolved them to form different kinds of creatures. The development level of the nervous system directly determines the grade of the organism.

Each creature and new life organization can be seen as an executable program. Programs can be combined, and code can be updated to develop more powerful programs. If this program works well and consistently reproduces, then it will survive, just as a program in a computer must be perfected to continue computing.

However, compared to the modern computers, the processing speed of this giant planet and biological computer earth is too slow; the operating process of a program is an organism's whole life. It cost billions of years to develop the highest level of intelligence—humans. Naturally, there are no other natural creatures on this planet that can evolve and have a better brain, except for the computers that are created by humans themselves.

The programs in the computer can be iterated quickly, but the artificial intelligence derived from this has not developed rapidly. Since rule-based programming and iteration are too much dependent on humans, the results are also constrained by us. What if we let the computer program by itself? After all, deep learning is based on the principle of nonlinear programming, allowing the program to transform itself. It is often very difficult to understand the logic deep-learning neural networks use to solve problems, just like humans do not understand how those ideas and thoughts in their brains are produced from brain cells.

The earth has long been covered with a biological layer. Today, this bio-computer ushered in a second evolution—that is, the layer of information is nurturing an evolution, with its surface wrapped by computers, communication networks, various types of sensors, and human activities. With the combination of data molecules and humans, new forms of data life are taking shape. Thus, they need a new brain.

Baidu Brain is such an attempt. If compared with the unconscious evolutionary journey of the biological world, then it is more concerned with the practical use at present: individuals, businesses, and society all are in urgent need of artificial intelligence, but artificial intelligence is still scattered everywhere. Baidu Brain has plans to provide concentrated and high-quality artificial intelligence that relies on the relevant information through the Internet neurons to accelerate the world's intelligence.

First Runner: Baidu Brain

A few years ago, Adam Coates was doing postdoctoral research at Stanford University. He asked his tutor Andrew Ng, "What do we do, and where do we do it, to make our research influence the world most?" Ng told him that he should go to Baidu. When talking about this anecdote, Adam had been the head of the Baidu's Silicon Valley AI Lab for more than a year.

Today, probably no one can deny the foresight contained in this dialogue. However, at the beginning, Baidu Research was a transfer station for Baidu employees who traveled to the United States.

In 2014, Baidu introduced the idea of Baidu Brain. This abstract concept only left a shadow on the media. After two years of work, Baidu "introduced Baidu Brain to the world for the first time" at the World Internet Conference at Wuzhen in 2016. At that point, the outside world came to know that Baidu had cooperated with more than thirty thousand companies.

The development of Baidu Brain was a hard job, but the R&D team members are not necessarily lacking in skill. In this team, some have played Subor games since childhood, some came from a town not covered by Baidu Map, some have regained their enthusiasm for the medical field while working on Baidu Medical Brain, and others have said, "It is hard and complicated," while insisting on studying how to "create the best experience."

Sometimes the benchmark of such extreme pursuit is even science fiction. Many young scientists in Baidu Brain love to watch sci-fi movies. Ordinary people see the genre as illusion. But this group of doctors and postdoctoral candidates see it as science. When watching the American drama *Westworld*, Gao Liang, senior director of Baidu's Speech Technology Department, said, "The plot makes me feel that the wake-up design, voiceprint recognition, and far-field technology are perfect. Yes, the future human-computer interaction should be like this!"

The construction of Baidu Brain also has a sci-fi element. Let's start by understanding its infrastructure. The basic layer of Baidu's artificial-intelligence business is the "material layer," which has a GPU/FPGA-based cloud-computing platform, a deep-learning code platform, and a big-data reserve. This layer provides an evolutionary environment and tool. The top layer of SaaS contains a variety of applications for artificial intelligence.

In between is the artificial-intelligence basic technology layer. The brain's cognitive functions such as "listening" and "speaking" (speech recognition and speech synthesis), "seeing" (visual recognition), and "reading and writing" (natural-language processing) are performed at this layer, and the decision-making functions that the brain possesses, such as deciding and planning, motion control and prediction recommendation, are also at this layer.

In a broad sense, Baidu Brain contains those three layers. Ya-Qin Zhang believes that the combination of these three layers reflects the comprehensive strength of Baidu Brain.

Baidu Brain is the core engine of Baidu cloud; Baidu cloud is the cloudification of Baidu Brain. The cloud provides the source of neuron and data training for Baidu Brain. Baidu Brain exports services to various industries through the cloud.

Specifically, at the material layer, Baidu is the first company in the world that uses GPU chips on a large scale in the field of artificial intelligence and deep learning and commercializes ARM (advanced RISC machine) servers widely. Baidu has also independently developed a server based on FPGA chips. Coupled with traditional CPU chip–based servers, integrating so many servers with different performance, structure, and principles requires powerful heterogeneous computing power. Through heterogeneous computing technology, 100G RDMA communication technology, and efficient whole-rack server technology, Baidu has built the world's largest GPU and FPGA hybrid heterogeneous computing cluster, combining hundreds of thousands of servers to form the entities of Baidu Brain, to ensure its superior computing power.

It's not enough to have a strong hardware structure; the brain needs content and data as well. This is like the memory of human beings. PaaS on top of IaaS is our artificial intelligence platform, with all samples, features, and functions growing at the PaaS layer. All the network data, search data, and billions of images, videos, positioning data collected by Baidu Search over the past decade become Baidu Brain's nutrients for its continuous learning and rapid development.

Apart from brain structure and memory content, Baidu Brain also needs to have thinking skills. The system simulates the neurons of the human brain through deep learning, which simulates the working mechanism through trillion-level parameters, hundreds of billions of samples, and hundreds of billions of features training. This is also the largest deep neural network in the world.

The artificial-intelligence basic technology layer includes speech recognition, image recognition, natural-language processing, and all knowledge maps, business logic, and user portraits.

The SaaS layer can be seen as the tentacles of Baidu Brain, which will be more vertical and penetrate into industries such as transportation, education, and finance. For Baidu, these three layers are the most capable services that Baidu Brain provides to business customers through the cloud, creating a smart environment.

Baidu Brain has completely surpassed the information-technology services of the past. For example, everybody used to do business by computing, storing, and networking. But, now the three layers are organically combined, and the soul lies in the artificial intelligence that penetrates the whole process.

Through the joint nourishment of hardware, data, and algorithms, Baidu Brain is growing like a rolling snowball and becomes increasingly efficient at processing data, extracting knowledge, understanding users, solving problems, and gaining more knowledge to realize the positive cycle of data→knowledge→user experience→new data.

Nicholas Negroponte, another Internet godfather, once said, "When I heard about the Baidu Brain, I thought these people are really crazy," making a machine's brain seems like a crazy fantasy in science fiction literature. But the beliefs and efforts of scientists make the seemingly "crazy" idea walk on the road toward smooth realization, just like the growth of a real life.

Training the "brain" is like the way a child learns: start from scratch, learn the language under the influence of a collection of writing rather than grammar, and form the impression of things from many pictures; this is the process to know the world through trial and error. Perhaps a one-year-old child can easily do something with less energy and time than Baidu Brain, which might require hundreds of times or even tens of thousands of times more energy and time. However, from another perspective, it is like a child of all human beings. It may inherit all the experiences and memories of the existing civilization. The evolution of the artificial-intelligence brain is essentially the evolution of human civilization with unlimited potential.

Baidu has opened PaddlePaddle, a free open-source, artificial-intelligence, deep-learning platform, and opened the Baidu Brain open platform at ai.baidu.com. The former provides an algorithmic programming environment for developers, and the latter provides a ready-made Baidu artificial-intelligence-results interface for application developers, data engineers, and data scientists. The openness of AI technology is the most efficient way to promote the industrialization of AI. The "AI energy" is continuously transmitted to various fields to bring about positive changes. At present, Baidu AI Open Platform has started more than two hundred core AI capabilities, with more than five hundred thousand access developers, covering various fields of real estate, corporate services, logistics, retailing, education, exhibitions, and smart communities. Baidu Brain is indeed sharing its resources and is also willing to melt the iceberg of artificial intelligence with all related companies.

Baidu Brain's Listening and Speaking Ability

The first step of human-machine dialogue is to let the machine learn to "listen" and "speak." Listening is the constant pursuit of accuracy, and speaking is to endow the brain with humanity.

One of the most basic abilities of Baidu Brain is speech-recognition technology. The function of listening has gone through several processes, i.e., from standard template matching to statistical model to deep neural network. Initially, the recognition of speech must go through the multistep transformation from acoustic model to the phoneme model and then to the language model. In recent years, under the training of a large number of corpora and deep learning, this step has been greatly simplified; the machine generates its own program from input to output, and the accuracy has been significantly improved. The Baidu Brain will "listen" more clearly.

In 2011, Baidu started working in the field of speech recognition. By 2016, the accuracy of Baidu's speech-recognition technology had reached up to 97 percent.

In the four years from 2012 to 2016, the accuracy of Baidu speech recognition increased by nearly 30 percent. Even for Mandarin with a serious local accent, Baidu speech can identify eighty-five out of one hundred sentences without any error. According to the standard of one wrong word, Baidu Speech Sound can accurately identify about 98 sentences, while untrained ordinary people can only understand about 60 sentences. In order to achieve dialect recognition, this recognition system needs to have corpus training for at least 720 hours. The sound, content, and speaker are constantly changed to improve the sensitivity of the system.

It is very difficult to let Baidu Brain "talk" in human language. Baidu uses acoustic models and language models. The acoustic model determines the pronunciation of the language. When a word is typed, the system looks for the appropriate pronunciation in the original sound bank. For the electronic pronunciation to sound human instead of machinelike, it is necessary to build a library for the speech sound material. For example, let the machine learn one hundred hours instead of twenty hours; after one hundred hours, the machine voice sounds much more comfortable. In order to maintain the continuity of the speech and avoid the brokenness of synthesized speech, the Baidu Brain language model will continue to improve the learning connection probability of the text library. For example, if you say, "China," the system can also select "People's Republic," "national," "sons and daughters," and so on in the follow-up phrases. (In Chinese, these words are often used together to form phrases.)

Long speech is a technology that makes machine speech more real. Emotional synthesis, far-field schemes, long-speech schemes, etc., can give emotion to synthetic speech, making it feel more like the human voice.

Speech recognition is extensively applied. For example, we can use the feature to create a strong salesperson. When a new salesperson calls his customer, Baidu Brain records the customer's reply in real time and displays it on the computer screen. The system can instantly search for and retrieve the responses made by the most excellent salespeople in the past. In this way, each novice can, on the first day of work, grasp the communication ability of the best salesperson in the past just by reading. In addition, during 2014, Baidu provided intelligent speech sound solutions for Tesla Motors. Chinese car owners can use their words to control car entertainment systems, direct map navigation, initiate searches, and even make calls via Bluetooth.

Accompanied by speech recognition, it can capture speech features. For example, after Hu Ge read the upper stanza of the poem "The Lantern Festival Night, to the Tune of Green Jade Table," the speech recognition system automatically generated a later stanza. The system can synthesize some distinctive star voices by recording and analyzing only about two thousand sentences.

At present, Baidu responds to about 250 million synthesized voice requests per day. After the speech-synthesis technology was launched, the daily duration of Baidu users who listen to voice novels increased from the previous 0.69 hours to 2.21 hours. This function gives people comfort when listening to novels. When elders and children at home miss each other, they can always be accompanied by the "voice" of the others.

Baidu's machine translation system, which is based on neural-network translation model technology, quickly learns various languages. After six years of accumulating information, today's Baidu's translation system can handle the translation of about twenty-eight popular languages around the world, covering 756 translation directions. Baidu's speech can recognize dialects such as Cantonese and the Shanghai dialect.

God once sectioned human language, so that people from different places could not communicate in different languages. With machine translation, humans can finally join hands and build a real Babel tower.[5]

Baidu Brain's Good Vision

Vision accounts for nearly 80 percent of our information-intake work. While solving the problem of listening and speaking, we should also teach computers to "see"; that is, image recognition. Consider a flower as an example. After the user uploads the flower image on Baidu, Baidu Brain converts it into a digital stream of zeros and ones and then inputs it into the deep neural network. Through layer-by-layer analysis and abstraction, the network compares the information of each layer, including the pixels with the existing big data, and the image is restored and recognized as a flower. This method is actually quite similar to the function of the human eye.

All of this must be based on preclassifying the images. At present, the world's largest image-recognition database, ImageNet, has more than one thousand categories. Baidu's image database has reached forty thousand categories.

Baidu is advancing the computer vision program in four areas. The first is face recognition. The face-expression network is formed for capturing the key points of a human face to realize accurate recognition. Second, for products that are similar to Baidu Map, the combination of map service with image-recognition technology creates an effect whereby data is infinitely close to the real world. In addition, Baidu's unmanned-driving technology also uses computer vision for program optimization, which accelerates the

---

5   According to Chapter 11 of the Old Testament's Book of Genesis, humans united to build a tower that could lead to heaven; to stop them, God makes humans speak different languages, making it impossible for them to communicate with each other. So the plan failed, and humans went their separate ways. This story attempts to explain the different languages and races in the world.

development of unmanned vehicles. Image recognition will also be applied to AR (augmented reality) to improve visual effects.

Baidu's face-recognition technology has far surpassed that of humans. There are more than two hundred million face photos in the Baidu database, and the number of images browsed during recognition training exceeds two million. At present, Baidu Brain can automatically determine the number of faces in the picture and the position and size of each face, supporting multiple perspectives such as front and side. Even if the target is in motion, the screening rate will not be reduced. The system can perform pixel-level face analysis by locating more than 70 key points such as eyes, eyebrows, nose, mouth, and cheek contours and identify people's gender, age, expression, posture, and other attributes based on face images.

Like the AlphaGo team, Baidu is also curious about where its technology boundaries are. Is there any deviation in the artificial intelligence R&D system? To this end, the Baidu team has participated in Jiangsu Television's large-scale scientific reality show *Super Brain* and competed against the "Water Brother" Wang Yuheng, who can distinguish 520 glasses of water with his naked eyes. In the previous program, "Water Brother" defeated the artificial intelligence Mark from Ant Financial. The Baidu team specially optimized the algorithm for the man-machine battle on-screen, which finally proved that Baidu team's robot Duer is really much better.

Among the products of Baidu's face-recognition technology, the most high-end one is the town Wuzhen's face-recognition access-control system. The facial information of those who have the access will be entered into the system in advance, so the people just need to scan their face when they enter any place where the gate is installed.

This technology is commonly called "one to one" in Baidu, which means a comparison of one face to the information in the database. Correspondingly, there is "one to N." This concept is often shown in Western spy films: the system searches for a person's facial information provided by the monitoring system to determine their direction in a boundless huge crowd. Although it is rather difficult to find this kind of smart skill in Chinese movies, in the "one to N" comparison, Baidu can really achieve more than 99 percent recognition accuracy.

This technique is easier said than done. When the technology matures and the databases are connected to each other, we won't need to show an ID card for inspection when we travel by plane or train. When we enter any transportation hub and the camera captures our facial information, the system can automatically confirm our identity and ticket information through face recognition. This means ordinary people can travel efficiently, and public order will be greatly improved.

Baidu's face-recognition system can complete the process of identification and judgment under the brightness equivalent of one candle in a space of more than one square meter. In the remote account-opening scenario, the biometric recognition technology can reach a response rate of up to twenty frames per second, and the interaction process takes less than two seconds. On this basis, combined with the video sequence, Baidu has applied face identification in the fields of financial antifraud, loan approval, remote ID-card identification, bank-card identification, etc., so as to accurately identify users and prevent fraud.

Baidu's "good vision" can also do a lot of other amazing things. When we took photos of the Hall of Supreme Harmony, or the Imperial Palace, from various angles, Baidu Brain removed overlapping and useless image information and completed the three-dimensional structure construction of the Hall of Supreme Harmony through calculation and modeling. In this way, people can immerse themselves in virtual tours through the Internet and feel the grandeur of the Hall of Supreme Harmony thousands of miles away. With more and more photos being uploaded, Baidu Brain can reconstruct more scenic spots, so that people around the world can experience 3D virtual tours without leaving their homes.

At the end of 2016, two passenger planes at Shanghai Hongqiao Airport almost collided on the ground—they were three seconds apart. The usual dispatching and warning functions of the tower did not play any role in this incident. Fortunately, the pilot did not wait for the tower command but engaged in emergency evasion, which avoided a major accident. This incident once again reminds us that omissions might be inevitable in the model that fully relies on the manual command of the tower.

Lin Yuanqing learned another such detail in the civil aviation department: to understand the road-surface conditions, the airport towers send staff to check the runway every four hours. This work is characterized by low professional demand, low salary, high labor intensity, and serious staff loss. It can be totally replaced by artificial intelligence: cameras can be installed near the apron to perform real-time 3D rendering of the runway environment combined with artificial-intelligence technology. The movements of airplanes, baggage carts, airport service vehicles, and all staff can be shown on the screen in real time. Additionally, parts and all foreign objects accidentally dropped on the runway can be found the first time without any omissions. The accuracy, predictability, and security of such systems are much higher than manual inspections.

The Era Calls for China Brain

During China's national Two Sessions in 2015, as a member of the Chinese People's Political Consultative Conference, I submitted a proposal for the establishment of the "China Brain Plan." The state will fund the building of the world's largest artificial-intelligence basic-resources and public-service platforms as soon as possible. The large-scale artificial-intelligence platform will house hundreds of thousands of servers; support data deployment; model debugging and application development of each plan participant; efficiently connect the intelligence, data, technology, and computing resources of the whole society; and completely rely on the unified platform to realize resource sharing and R&D innovation promotion. This will boost the new round of the industrial revolution. The results of basic research should benefit more Chinese companies. Speech recognition, image recognition, natural-language understanding, multilingual translation, and even robots in unmanned vehicles and smart manufacturing can perform a variety of innovations and practices on this platform.

If the plan is only implemented by Baidu, it may only provide tens of thousands of servers, but if it is led by the government, then we will have hundreds of thousands of servers. A large platform can reduce cost and encourage innovation. The country's sustainable and steady large-scale investment has allowed a large number of enterprises to grow up, and more and more innovations have followed, laying the foundation for China's position in global innovation in the next ten or twenty years, or even longer. This is my long-standing idea: I don't care how Wall Street thinks, I must make this happen.

However, any super engineering may be subject to controversy.

In September 2016, a high-energy physics event unexpectedly set off an argument on super-large particle colliders that extended from academic circles to society. Ordinary people have begun to pay attention to the term *particle collider*.

The importance of particle research is vividly expressed in the science fiction novel *The Three-Body Problem*. To prevent the science and technology advancement on earth, the alien intelligent creatures use the principle of quantum entanglement to create a Sophon, with an eleven-dimensional form, and launches it on the earth. The Sophon, which moves at the speed of light, can simultaneously interfere with all particle colliders, destroy the collision results, lock down the basic physics research, and confine science and technology at a lower level.

The argument began with the debate of two outsiders, American mathematician Shing-Tung Yau and Harvard University PhD physicist Mengyuan Wang, which lasted for three months. It also disturbed Yang Zhenning, winner of the Nobel Prize in Physics, and Wang Yifang, director of the Institute of High Energy Physics of the Chinese Academy of Sciences.

In the face of the debate over whether to build a super-large particle collider in China, opponents pointed out that the construction of the collider would cost hundreds of billions of dollars and the power consumption would be comparable to that of a big city, but the projected results seem extremely uncertain. The collider might become a big toy for the physicists. The United States has shelved a similar plan, and the achievements of the European Large Hadron Collider[6] were few. So why would China create a collider?

Proponents argued that it is important to study the Higgs boson[7]—the God particle—which will unravel the creation of the universe. The rejection and delay from the United States and Europe had just provided an opportunity to China. As a rising power, China should assume the responsibility of cutting-edge research in theoretical physics.

---

6   The Large Hadron Collider (LHC) is the world's largest and most energetic particle accelerator. It is a high-energy physical device that accelerates the collision of protons.

7   The Higgs boson is a spin-zero boson predicted by the standard model of particle physics. The physicist Higgs proposed the Higgs mechanism, in which, the Higgs field causes spontaneous symmetry breaking and imparts quality to the canonical propagator and fermion. The Higgs particle is the field quantization excitation of the Higgs field, which obtains quality by self-interaction.

In the end, the debate was suspended because the super particle collider was not approved by the thirteenth Five-Year Plan.

For more than five decades since the founding of New China, big plans had come one after another. Since the first mention of *smart computer* in the 863 Program from the early days of reform and opening up, Chinese scientists have been catching up to advanced countries for over thirty years. A rarely known fact is that in the era before supercomputers and big data were invented, Chinese artificial intelligence started from sound-recording libraries of university laboratories. Thanks to the persistence of these predecessor scientists, China's economic, scientific, and technological strength all are increased day by day, and Internet companies can grow rapidly in China.

At present in the Baidu artificial-intelligence research labs in China and the United States, more than 1,300 researchers of different ethnicities and nationalities are working on hundreds of related projects day and night. Their achievements will all be transferred to Baidu Brain. These developers are like the over one thousand scientists in the Manhattan Project, or the over three thousand researchers at the European Nuclear Research Center today, doing things that are ahead of the times and not yet understood by people. With improved algorithms, upgraded modeling, and analytical processing, Baidu Brain researchers, like the particles flying in particle accelerators, are sparking a revolution in intelligence.

The China Brain Plan is somewhat different than the super large particle collider, which covers a huge area and consumes a lot of energy (the European particle collider needs 12 million kilowatts of electricity). The development of the China Brain is a bottom-up and natural project. The country does not need to spare effort. All we need is direction and determination.

Economy of scale is the basic factor for the success of Chinese industry. In our big country with more than 1.3 billion people, about 700 million netizens, and tens of millions of engineers and scientists, the massive data, abundant talent, rich business, and various application scenarios are like running water. If we can't make the best use of it, then we will surely miss this wave of intelligence, yet it will be a real waste if we give up the technology's commanding heights and national security. As massive amounts of data on the Internet have spawned stream data-processing technologies such as

Hadoop (a distributed system infrastructure developed by the Apache Software Foundation) and Spark (also from Apache, it is a common parallel framework like Hadoop MapReduce, which is open sourced by the AMP lab at the University of California, Berkeley), still artificial intelligence has been scattered throughout China, like nerve nodes, promoting the arrival of China Brain with its brain waves—this is the call of the times.

China Brain, China Style

Fei-Fei Li, the codirector of Stanford's Human-Centered AI Institute and the founder of ImageNet, the world's leading image-recognition database, said, "From science to technology to products, it is like a 4×100 relay race. Every stick has its special function; the academic field should be regarded as the first stick of this 4×100 relay race, the industry and the laboratory being the second, and investment is the third and fourth stick."

As an example of the most critical and exciting last two sticks, in January 2013, the European Union announced an investment of about 1 billion euros to simulate the entire human brain with a giant neural-network computer. In early October 2016, the United States had issued two reports, *National Artificial Intelligence Research and Development Strategy Plan* and *Preparing for Future Artificial Intelligence*, to formulate artificial-intelligence research and development strategy planning. In the same year, artificial intelligence was frequently mentioned in China's official reports. In May 2016, the *Internet+ Artificial Intelligence Three-Year Action Implementation Plan* was jointly issued by the National Development and Reform Commission, the Ministry of Science and Technology, the Ministry of Industry and Information Technology, and the Central Network Office. In August 2016, the State Council's "Thirteenth Five-Year Plan on National Science and Technology Innovation" was released, and artificial intelligence once again became the core hot point.

Will artificial intelligence be universal? Of course. This is the main point, but this universality is not to be provided by a single country. The development of artificial intelligence must be as diversely competitive and local as organic evolution. It is worth pondering that when the US tech giants started to compete on the artificial-intelligence track, they unexpectedly suffered from acclimatization in China.

CHINA BRAIN PROJECT: BOTTOM-UP SUPER ENGINEERING

At the end of 2016, IBM's medical robot Watson arrived in Shandong. This highly successful robot in the United States was caught in a language problem when it came to "work" in China. IBM lacked sufficient Chinese data, which made the cooperation crack from the beginning. Watson, the "technical god," can understand many languages but can't do anything about the Shanghai dialect, Cantonese, and the southern Fujian dialect.

This is not an isolated case. The life, thinking, and cultural differences between the East and the West are not often addressed adequately.

The searching focuses of Chinese and foreign netizens are totally different. In Google's 2015 Hot Searches list, the first is NBA star Lamar Odom, the second is the woman fighter Ronda Rousey, and the third is US TV show celebrity Caitlyn Jenner. Very few Chinese will search for these people. However, in the 2015 Baidu hot-search list, the top searches are for Jin Xing and Wang Sicong, which seems more down-to-earth for Chinese.

Microsoft chat bot Tay played on Twitter for twenty-four hours and learned about swearing, racial discrimination, and gender discrimination, which is very "American."

Obviously, in a diversified Internet world, no country or institution can take charge. Only when those of different cultures, economics, and political backgrounds cooperate can they make a comprehensive and appropriate response to the needs of netizens.

From the perspective of national conditions, China has a stronger internal drive than European and American countries toward the development of artificial intelligence. This drive comes from the people. According to Baidu search statistics, the number of search requests for "services" has been growing rapidly: In 2014 it was 133 percent more than in 2013. In 2016, with a larger base, there was still a 153 percent increase.

China's use of mobile Internet is greater than that of the United States. Chinese netizens have long been accustomed to resorting to the Internet for services. Today, fifty-five out of every one hundred movie tickets are ordered through the Internet in China. Correspondingly, the penetration rate of Internet ticket purchases in the US film industry is only 20 percent. When the Internet penetration rate of the Chinese catering industry is 2 percent, the US penetration rate is only 1 percent.

The greatness and specialty of China makes artificial-intelligence innovation the only way to effectively deliver service needs to Chinese netizens.

Industrial anxiety will also drive the usage of artificial intelligence. The cost of manufacturing in China is rising rapidly.

The global industry is becoming more and more automated and intelligent. High-end manufacturing may return to Europe and the United States, and low-end manufacturing has begun to flow to countries such as Vietnam. If industrial transformation is not completed in a short period, then China's manufacturing industry will face "hollowing": high-end and low-end manufacturing will both flow out of China. Can this transformation be done without artificial intelligence?

Despite the urgency over industry, China's strength is worth our optimism. The executive power of Chinese companies and the support of the Chinese government are strong backup forces for the emerging industries.

If the needs of netizens dovetail with the cooperation between enterprises and the government, then data will be essential for the development of the China Brain. In this field, China is unique.

The large population, complex social environment, and different application scenarios of Internet companies are all significant for the data collection. It is foreseeable that in the near future, in addition to personal data, data generated and accumulated based on the public environment or the government, such as automobile-registration information, academic qualifications, criminal records, etc., will form an encrypted personal basic electronic file. The data obtained by the market through the service output, such as credit-card bills, consumption records, website browsing preferences, and custom mobile-phone brands, will be converted into service feedback of the user again in the form of user authorization.

The users here are not "code farmers" who can code and make models, but ordinary public groups. It is truly a technological benefit to make it easier for more people to use smart devices.

The US Apollo program caused a large number of different enterprises to grow and innovate. The ARPANET used by US military during the Cold War (the United States Department of Defense's Advanced Research Projects Agency Network) delivered the Internet. The contribution of the China Brain, an imaginative super-engineering project, for the Chinese economy is not only the emergence of a number of star technology companies and scientific and technological achievements but also services and impetus for the transformation of the entire economy and society.

CHINA BRAIN PROJECT: BOTTOM-UP SUPER ENGINEERING

In 2017, Baidu was given the approval to build the National Engineering Laboratory of Deep Learning Technology and Application. Members of the building team include Lin Yuanqing, former Dean of Baidu Research; Xu Wei, distinguished scientist of Baidu Deep Learning Laboratory; academician Zhang Wei of Tsinghua University; and academician Li Wei of Beijing University of Aeronautics and Astronautics. Baidu will work with Tsinghua University, Beijing University of Aeronautics and Astronautics, China Institute of Information and Communications, China Electronics Standardization Institute, and other organizations to integrate the company's advantageous resources and build a "domestic-leading but world-class" deep-learning technology and application-research institution, for enhancing the overall competitiveness of China's artificial-intelligence field through research breakthroughs, industrial cooperation, transfer of technological achievements, and personnel training.

This will be the beginning of Genesis's super engineering.

The humanist painter Michelangelo completed the majestic *Genesis Fresco* on the ceiling of the Sistine Chapel, in which the hand of God touches Adam's fingers and the enlightenment of wisdom arises. In this mural, God's robe is wide and flowing. In recent decades, it has been pointed out that the shape of the robe of God is actually an anatomical picture of the human brain. In this mural, Michelangelo quietly hid the enlightenment code of God in the human brain, and human beings have enlightened themselves!

Human beings develop their own wisdom in self-work and self-enlightenment. Today, deep-learning neural networks also create a new "brain" that is self-operating and self-tuning. This huge brain of artificial intelligence will become a new background for human civilization; it will be the embodiment of the great activities of mankind and support human civilization to take steps forward.

5

CHINA'S INTELLECTUAL MANUFACTURING AND CIVILIZATION UPGRADE

One day in late 2016, when Baidu Map held a press conference and announced that its daily location-based service exceeded 72 billion accesses, Li Dongmin, former general manager of Baidu Map Division, couldn't help but think about the first time he heard Lu Benfu's speech on Internet economy over ten years ago. At that time, he was in graduate school; he said he "saw the light in the confusion" and soon seized an opportunity to work as an intern at the newborn Baidu.

In 2000, Lu Benfu, an online celebrity scholar at the time, talked about the newborn Internet everywhere. In a forum, when questioned by a listed company CEO, he replied, "You are a successful person. Why did you succeed? Because during the economic downturn during [the] 1970s, you were the first one to focus on the quality; in the market economy of the early 1980s, you were the first one to engage in brand marketing; in the surplus economy of the 1990s, you were the first one to engage in chain store and company operation. You succeed because you always stand at the forefront of the times. But, what do you know about the trend after 2000?" The boss seemed to be enlightened, and immediately said he would invest in the Internet.

With the rise of the Internet economy, the Internet thought market has also been unprecedentedly prosperous. Famous and unknown Internet-wealth propagandists have been talking eloquently, and Lu Benfu's[8] views

---

8   Lu Benfu, a Chinese scholar on Internet economy.

have long been obsolete. However, how many of these ideas can dominate the next wave?

Artificial intelligence is comparatively new, so to discuss its impact on society, it is not enough to solely consider it relative to the Internet economy. This is the most cutting-edge change in the long process of social evolution. Not only will the overall economic and technological conditions change, but also the national social-governance level and even the cultural and individual levels will change because of the infiltration of artificial intelligence. The changes in the economic base and superstructure will affect the way in which civilization is made. Can artificial intelligence help to realize the ideal of universal society? Can it contribute to a harmonious yet different social order? To answer these questions, we need hard work and imagination.

From Hardworking Revolution to Intelligent Revolution

Scholars such as Fei Xiaotong, Kaoru Sugihara, and Giovanni Arrighi have described a path that can be called a "hardworking revolution" in China's modern history of rejuvenation. The hardworking revolution is totally different than the heavy investment of the Western industrial revolution. It is characterized by relatively cheap but basic education of skilled labor in relatively small-scale units, labor-intensive industries, and the ethics of hard work and wealth. This road is not the opposite of the industrial revolution but instead a reply to the impact of the industrial revolution. Through hardworking revolution, China has completed the counterattack on the industrial era under the condition of poverty become the world's number-one manufacturing country, with a rare global industrial chain and the number-one GDP growth rate in the world.

Over the past three decades, China has grabbed almost every opportunity. During the third industrial revolution, China initially completed industrialization and took steps that made up for the loss of the prior century. During the 1980s, when Western countries pursued deindustrialization, China undertook the industrial transfer and became a world factory. In the late 1990s, when the Internet took off, China finally stood on the same starting line as the United States. Nowadays, China's development of mobile Internet is even slightly better. The glory of China's reemergence is partly due to globalization. Joining the World Trade Organization (WTO) made Chinese goods easier to be sold on the global market. In various industrial fields, Chinese people learned advanced industrial technologies from advanced countries very quickly. When Chinese people learn how to manufacture something, the price of it is reduced. Therefore, China will often be referred to as a free rider. But now, when China has become the world's second-largest economy, more and more people think that China can no longer fully carry on the long-term following strategy. China also needs to take the leader's responsibility, not only to create economical high-speed trains for the world but also to provide technical guidance and even a model of civilization.

Yet, the hardworking revolution has encountered a crisis. The aging population, rise in labor costs, big-city diseases, anxiety of the emerging middle class about environmental pollution, increased international competition, and dissatisfaction caused by the widening gap between the rich and poor all plague the country. If the direction is wrong, working hard won't help to lead the trend. The Chinese people are eager to find a new high-speed development path. If the direction is clear, they are willing to be hardworking again to learn and catch up. Viewed from the industry perspective, the entire country needs to eliminate and transform backward production capacity, upgrading the industry.

In terms of upgrading, some public opinion is that China lacks innovation and can only focus on low-end industries, but, in reality, this is not the case. The expression of the hardworking revolution itself has somewhat masked Chinese people's ability for technological revolution. China's scientific and technological strength has been greatly improved, and even in the high-tech areas, it is in one of the few advanced countries. In August 2015, Dr. Yuan Lanfeng from the University of Science and Technology of China wrote an article that argued, with information from various sources, such as the *Nature* magazine index, data of the world's top five patent offices, scientific-research input, and the status of advanced projects, that China's overall scientific and technological level is approaching the number-one country, the United States, and is currently ranked second in the world. This article was forwarded widely and triggered a hot discussion in scientific and technological circles.

In October 2016, the Chinese Academy of Sciences and Clarivate Analytics (formerly Thomson Reuters Intellectual Property and Technology Division) jointly released *2016 Research Frontier* based on the paper's big-data analysis.

It said, "China has a large gap with the United States in terms of cutting-edge guidance, and its competition with the UK is fierce. In terms of potential leadership, China has surpassed the UK and ranks second in the world, showing strong follow-up development capabilities. China has participated in 68 frontier directions, 30 of which lead the world. There is still a big gap with the No.1 US, but it has already surpassed in many areas."

In terms of industry, China's large technological enterprises have emerged, and top companies in engineering-machinery manufacturing and telecommunications can compete with multinational giants on a global scale. Although there has been great progress in the transformation from labor-intensive industries to technology and capital-intensive industries, the gap between China and advanced countries is still obvious. Taking industrial robots as an example, according to statistics provided by the International Federation of Robotics (IFR), in 2013, the number of robots in China's manufacturing industry was only twenty-five units per ten thousand people, while the world average was fifty-eight units per ten thousand people. South Korea had 396 units per ten thousand, Japan had 332, and Germany had 273. For the automotive industry, where most robots are used, the density of industrial robots in advanced automobile-producing countries has reached one thousand units per ten thousand people; in China it is only 213. However, since 2013, China's industrial robot sales have accounted for 20.52 percent of the world's total, surpassing Japan for the first time to become the world's largest industrial-robot sales country. According to data of 2016, the density of industrial robots in China has reached forty-nine units per ten thousand people.

China's rapid advancement in the field of artificial intelligence has attracted worldwide attention. According to the *Washington Post*, China has published more research papers in the field of deep learning than the United States, with high quality as well. China is also active in the field of application. In the 2017 International CES (Consumer Electronics Show), the number of Chinese companies has exceeded one-third of the total number of exhibitors, and many Chinese artificial-intelligence products won the official Best Innovation Product Award.

In the United States, the capital-initiated high-tech industry had developed rapidly during the Obama administration and has transmitted technology to the world, but it has also increased its division. The hollowing of the manufacturing industry makes high tech such as artificial intelligence unable to fully put into practice in the country. In China, the large-scale manufacturing and service industries, coupled with the millions of engineer graduates every year, has facilitated the use of artificial intelligence.

The First Manufacturing Power under the Impact of Three Technological Waves

Anxiety existed for many years. In the late 1990s, a jingle was popular in the industry: "Now the refrigerator can freeze, the color TV has an image, the PC can be customized, and nobody really understands ERP." These words are very concise. The first two phrases reflect the predicament of industrialization. Home appliances are the iconic industries of "made in China," which have experienced blowouts in consumption and production capacity. Enterprises had been plagued by quality problems in the early days. A few companies are committed to improving technology and management and stick out from the crowd. Although standardized functional requirements such as "freeze" and "has image" tend to be realized, high quality means having a long life cycle and low replacement rate. Here, the market is stagnant.

Although the PC can be customized, it is still limited to CPU frequency, memory, hard-disk capacity, and other performance, which is a fairly basic customization determined by the modular structure of the traditional PC. It is incomparable with the advanced smartphone that supports applications and content customization, data production, sharing, and feedback. In the future, artificial-intelligence terminals will be far more open than smartphones.

ERP (enterprise resource planning) was once seen as the core of the enterprise information system, and now it is declining. The ERP template development is equivalent to conducting manual deep learning on a large amount of enterprise data and finding out the basic model. It is also possible to carry out secondary development for each user's characteristics. But in essence, it is still a mode of centralized development, with a lack of flexibility, long deployment cycle, and disconnection from the business, which is the root cause of "no one understands ERP." ERP can't keep up with the changes happening in the business environment in recent years, and it is gradually being replaced by cloud computing. Cloud computing is not just a technology, but it also represents an idea. As with cloud (distributed) computing, enterprises must shift from traditional centralized management to distributed management.

At the macro level, manufacturing is marginalizing. In 1982, John Naisbitt wrote the following in *Megatrends*: "Japan has replaced the United States as the world's industrial leader. . . . Japan is the champion, but only a new world champion in a recession game." Naisbitt believed Japan would also be challenged by some emerging markets. At that time, China was in the early stage of reform and opening up and had not yet entered his vision. Twenty years later, China has become the latest and possibly the last world champion in the industrialization competition.

Just over three hundred years ago, the vast majority of the world's population was farming in the fields. No one could imagine that people would soon flood into cities and factories. In the United States today, the number of small farmers is less than 1 percent of the working population. More than a century ago, a similar scene was staged at a higher level; at that time, people began flooding into office buildings. More than seventy years ago, the number of white-collar workers in the United States exceeded the blue-collar workers. More than twenty years ago, the US business community abolished a large number of white-collar jobs, and many people left the nine-to-five office voluntarily. The organizational model of traditional industries is undergoing tremendous changes.

It is often said that the United States of today is China's tomorrow. China is also experiencing the cycle of production model changes in a much shorter period. More than thirty years ago, most of China's population was working on the farm. More than twenty years ago, the white-collar class appeared in China. In recent years, the career prospects of white-collar workers have reached a certain bottleneck.

This became a new law of imbalance. The expansion in the United States occurred simultaneously in China. As more and more people started their own Internet companies, a large number of new factories and new offices were put into operation, and lots of migrant workers, blue-collar workers, and white-collar workers were recruited. This situation is complicated, and it is a more challenging test for the wisdom of the whole country.

Another futurist, Alvin Toffler, author of *The Third Wave*, commented on a trip to China in 2001: China contains three worlds, the first wave covering about nine hundred million farmers, the second wave covering about three hundred million investors, and the third wave, according to data from the National Planning Commission (later reorganized as the National Development and Reform Commission) that Toffler received at the time, was only ten million upper-level managers. The theme of China's development is to change the relationship among the population in the three waves.

Today, the demographic pattern has changed drastically. According to data released by the Chinese government, the urbanization rate of China in 2016 had reached 57.35 percent. The resident population in cities and towns reached 770 million. During the "Twelfth Five-Year Plan" period, the urbanization rate increased by 1.23 percent per year, and urban population increased by twenty million per year.

Although the third wave had created many jobs, its border with the second wave was vague. The Internet industry itself includes the three worlds. Consider Baidu Takeout as an example. The company has a large number of takeout delivery staff, which is equivalent to the traditional blue-collar work. The operation and maintenance personnel are equivalent to the traditional white-collar workers. There is also a small and sophisticated technical department, including the artificial-intelligence team, which is equivalent to the top level. As far as the company is concerned, it is not a small challenge to manage the three groups of employees with such big disparity in culture. From the outside view, this is the characteristic of China's development path, and we need to balance this contrast and mix.

China's transition from the first wave to the second wave already passed the Lewis Turning Point in approximately 2010.[9]

Some of China's low-end manufacturing industries have moved to those countries with lower labor costs, such as Vietnam. India is also ambitiously planning to become a global manufacturing center. In September 2014, the Narendra Modi government announced a new Indian manufacturing policy to the world, including plans to provide one-stop services, to reform labor

---

9 Arthur Lewis, Nobel Prize winner in economics and a leader in development economics, analyzes the dual economy of developing countries. He believes that when the rural surplus labor force is exhausted and the urban and rural areas form a unified labor market, wage levels will continue to rise.

laws and taxation, and to simplify the approval process to attract countries around the globe to invest in India to set up factories and increase local employment opportunities. India's multitemporal and multispatial mixed traits are more serious than China. English-speaking intellectualism and the backward caste system are contradictory. We can wait and see whether India can become a strong opponent of China in manufacturing.

In addition, traditional manufacturing powers such as Europe and the United States are working hard and hope to go further. The German Federal Ministry of Education and Research and the Federal Ministry of Economic Affairs and Technology came up with the concept of Industry 4.0 (the fourth industrial revolution) at the Hannover Messe in April 2013, which was widely recognized by the government, industry, and academia. Industry 4.0 has risen to Germany's national strategy and triggered a new round of global industrial competition. Industry 4.0 aims to improve the intelligence level of manufacturing and use the cyber-physical systems (CPS) to digitize and intellectualize the information of supply, manufacturing, and sales in the producing process and to integrate customers and partners in the value process, to achieve fast, efficient, and customized product supply.

The United States is trying to reverse the trend of post-war deindustrialization, helping some factories return to the mainland. Because the country is big, the situation is complicated, and various powers are scheming. For example, unmanned-vehicle factories such as those belonging to Tesla Motors do not help employment, the income level of the middle and lower class in the United States has shrunk, many old factories have not returned to the Rust Belt[10] but are situated in the South where labor cost is lower, etc. President Trump continued to lambaste Toyota, GM, and Ford on Twitter, threatening them to move their factories back to the United States, which highlights the weirdness of a space-time disorder. But in any case, the US's actions to revive the manufacturing industry are bound to put pressure on China's manufacturing industry.

The main innovation of the US manufacturing industry is 3D printing. Although it appeared in the form of futuristic technology, the key to 3D

---

10 Rust Belt originally referred to the area near the Great Lakes in the Midwest of the United States, where traditional industries suffered due to a recession. Now it refers to areas of industrial decline.

printing is not the technology but the transformative trend of manufacturing from large-scale standardized production to mass customization. It finally reflects the trend of personalized and self-organized customer demand.

In the face of such pressure, China's manufacturing industry is in a precarious position. Labor costs in the United States can still be transferred to the Midwest. China as a whole is gradually losing low-cost advantages, and there is an urgent need there to shift competitiveness to productivity and a knowledge economy.

Ya-Qin Zhang often communicates with companies about cooperation on the Baidu cloud-computing business. He believes that China is far behind the United States in terms of industrial Internet and smart manufacturing. Over the years, most of the major industries in the United States have been streamlined. Most companies in vertical industries have used ERP, which cost trillions of dollars to make the process facilitated by IT and software. Automation starts very early in manufacturing, including high-precision machinery and assembly lines. Overall, China is relatively backward in this respect. Chinese companies, especially many small ones, lack a unified workflow and IT penetration.

The transition from the second wave to the third is just starting, even in the most advanced United States. To some extent, Silicon Valley's innovation is mainly about technology, relatively independent of the industry born in the second wave, and the mutual integration may not be as good as in China. Kevin Ashton, the originator of the concept "Internet of things," is extremely optimistic about the development of China's Internet and Internet of things. Ya-Qin Zhang believes that when the consumer Internet emerged, all the countries were on the same starting line. The United States can succed, and China can too, maybe even better. Now artificial intelligence is emerging; the original advantage is less important and may even be a disadvantage. For example, traditional companies spend a lot of money to buy a database management system from Oracle, and ERP and a database management system from IBM. Over the years, companies that are based on traditional IT technology have become more and more professional, their hardware and software turning increasingly complex. IT investment, operating costs, and labor costs are getting higher, so many companies are overwhelmed. Cloud computing is a new equal starting line. The cloud removes the complexity and lets the cloud company do the IT part for the enterprise. Unlike the traditional ICP (Internet content provider), service companies, which send hundreds of people working in the client company, provide virtualization service. For example, in the past, we built power stations and water wells in our own houses. Now, the electricity grids and water networks are provided so that enterprises can get water by turning on the tap. They do not need to deploy so many servers or such heavy ERP systems either; instead, if they can get access to a network, all services can be connected—computing, storage, database. Therefore, the commercial end service is also consumerized, which is convenient and fast.

Previous Manufacturing Power: Man Driven by Things

With the upsurge of the artificial-intelligence industry, optimistic and pessimistic views once again abound. When artificial intelligence of the twentieth century was just formed out of imagination and theory, the controversy began. People's fears and hopes reflect the tangled relationship between people and technology that runs through the three waves. A brief review of the history of technology and tools helps us understand the relationship between manufacturing and artificial intelligence.

The customization of goods is supported by the history of biological evolution. Evolution tells us that creatures adapt to the environment by changing traits. This iterative process is very slow, and acquired knowledge cannot be written to the program (DNA). However, in terms of tools, human beings can transcend the evolution of physiology, integrating acquired knowledge and skills through the improvement and iteration of tools, thus transforming the environment and changing themselves.

Manufacturing tools have even become the essential definition of human beings. Although some animals use tools—such as human's close relative, the primate—the tools used by these animals are still conditional. The human tool-making ability is not just a response to immediate challenges but also a deep learning from past experiences and prediction for the future. For instance, consider the stone axe with the multifunctions of cutting and peeling, which is suitable for different situations and became the tool our ancestors carried.

Artificial intelligence has just entered its own stone age. If computers can only complete the task they are programmed for, then human beings are committed to teach it to develop "tools" in the future.

Let us use the case of the automobile manufacturing industry for explanation. Frederick Winslow Taylor, a follower of the scientific-management school of thought, went through trial and error (deep learning) on the work of loading and unloading and concluded that the optimal load of a single person is 21.5 lbs. (9.75 kg), so a special shovel was designed for each worker to suit this capacity. As each worker differs in strength and endurance, this optimal weight reflects the average physical capacity of the workers in Taylor's time. Although the shovel is a tool for workers, the manager sees the standardized shovel as a way to make all the workers conform. Single workers are the flesh tool of the corporation, and workers become a tool for the shovel. That is exactly "man driven by things."

In the early days of the automotive industry, each worker independently assembled a vehicle in a way he was most accustomed to. After a long period of study, workers can reach a certain level of proficiency, but the method is not the most efficient. The next step of evolution is the assembly line.

It is said that one day Ford executives visited the slaughterhouse and got inspired to assemble the cars on assembly lines, in which the car was divided into the smallest and simplest unit and each worker was responsible for only one process. In 1913, Ford developed the first assembly line in the world. The chassis and parts of cars were sent to the workers through the conveyor belt and the workers did not have to walk around, saving the time for fetching parts. Previously it took 728 man-hours to produce a car by hand, but the assembly line shortened it to less than 12.5 man-hours. Car prices fell sharply, becoming affordable for civilians. The Model-T was put into production in 1908, and the total output reached 15 million in 1927.

The early automobile assembly lines contained 7,852 kinds of jobs, of which 949 required good physical strength, 3,338 only required normal strength, and 3,595 needed below normal physical strength. Of the jobs needing only below-normal strength, 715 could be completed by a one-armed worker, 2,637 could be completed by a single-legged worker, 10 could be done by blind people, and 2 could even be done by those with disabled hands. This detailed division of standards can be understood as a more primitive characterization and data breakdown, which implies that today's data-tagging technology and tagged data are the fuel of artificial intelligence, which is still a product of the industrial age and unable to break from chains of man driven by things or to meet the needs of humans at a higher level. It will first encounter the bottleneck of diversity.

Only New Industrial Automation Can Match Human Diversity

Diversity is a fundamental feature of living things. The diversity of common organisms is generally demonstrated in different species; that is, the traits of the same species tend to be the same. But in the advanced stage of evolution—intelligent life—diversity is manifested among individuals, especially in organs such as brains and hands, which are associated with wisdom.

We can imagine rolls of dice in the human gene that randomly generates different DNA codes. Even nontwin brothers and sisters have significantly different talents. People can always feel the differences with each other, but limited language makes it hard to fully express these differences. In the past, people have come up with many personality classifications, constellation being the most popular, which show people's desire to express distinctions. The distinctions of individual traits of the organism confused Darwin, but they have laid the biological foundation for the division of labor and the diversification of consumption in an economy.

How does traditional industry mesh with the needs of diversity? Henry Ford, the industrial pioneer, said three sentences that show the relationship between business and consumers in the old industrial paradigm: "If I had to ask customers what they want, they will tell me: a faster horse." "Any customer can have a car painted any color that he wants so long as it is black." "Every time I reduce the charge of our car by one dollar, I get a thousand new buyers."

What are the specific requirements of cars users? Xue Cheng once wrote an article in *China Entrepreneur* to analyze this issue. Before the advent of the car, except for a few engineers, the general public knew nothing about it. Engineers could only increase the rate (faster) with known demand (horse). According to Maslow's theory of demand hierarchy, the lower the level of demand, the easier it is to quantify, the greater the commonality becomes, and the larger the corresponding market size is. Adam Smith pointed out in *The Wealth of Nations* that scale is the premise of division of labor and cooperation. If we want to increase product, more advanced demand means more complicated division of labor as well as lower efficiency, and scale effect will be lost. For automotive production, the first task is to expand production for easily satisfied needs. "Each time we drop the car price by one dollar, we will have a thousand more customers." If only two hundred customers are added for every dollar invested to increase the color value, eight hundred potential customers will be lost. So, Ford only offered black cars, but that gave priority to increasing the production and reducing costs.

But at the consumer level, when car customers are created, their new demands are also created. Peter Drucker's *The Practice of Management* clearly states that the purpose of the company is to create customers. When productivity expands, consumer demand will be released or created.

Frederick Winslow Taylor believed that employees work for money, and Ford also said, "Wage solves nine-tenths of mental problems." But from 1924 to 1932, psychologists (rather than management scientists) headed by Elton Mayo, through experiments at the Hawthorne plant, found that worker's work motivations are much more complicated. In 1960, psychologist Douglas McGregor proposed theory X and theory Y. Under the former theory, managers regard workers as lazy and in need of being motivated by whip and reward. Managers adopting the latter theory believe that people have initiative and creativity.

The entire old industrial economy is like the solar system, where consumers and employees run around the enterprise, and the business runs around the finances (the sun). The system responds to the diverse needs of people but operates in a centralized manner.

Technology is still evolving; unless new species are created, the direction of product diversification is approaching the limit of human perception. A new cycle will be formed. The new industrial paradigm will reverse the old industrial paradigm. A gradual change in demand caused by sudden change in technology will be replaced with an infinite expansion of the level and variety of demand, with a gradual change in technology resulting from a sudden change in demand. Expanding domestic demand should not be regarded as an expansion at the old demand level but an unprecedented development of diversified needs. Under the old industrial paradigm, we could only rely on the reference system of known technology and known needs. Today, human's segmentation needs of infinite diversity rely on artificial-intelligence deep-learning systems.

The dawn is big data and artificial intelligence. Today, human diversified needs and feedback are increasingly being recorded by sensors and digitized and can be generated indefinitely. Only deep learning and intelligent economic systems based on probabilistic and distributed computing methods can sense the future direction from this infinite data.

The example of cars can clarify this issue. Marshall Fisher from Warton Business School once visited car dealers to learn about the color, interior, engine, and other functions; with all combinations, car manufacturers can actually provide twenty million models. Customization takes eight weeks, but more than 90 percent of the customers want to buy from stock, being unaware of so many possible models. The dealer has only two models in stock, and there are ten dealers in the local area. Assuming the dealers are of the same scale, only twenty models can be offered in the local market. Fisher therefore compares that to the neck of an hourglass.

The survey was published in the *Harvard Business Review* in 1997. During the same year, Amazon was launched, providing a solution—the network shelf is infinite. The service provides a matching function, which records and analyzes users' behaviors and speculates on their interests, which is the so-called user portrait. With this method, an unlimited number of product models can be matched accurately with an unlimited number of user needs.

Yu Liang, from the China Research Institute of Fudan University, wrote an article in Huxiu.com, in which he stated that many people misunderstand user portraits to be a description of groups of users. On the contrary, the user portrait based on artificial intelligence is just a description of the individual. It can associate any number of labels with everyone and carefully track each individual's needs. For example, all applications that recommend information by algorithms are actually user personality collectors that are disguised to push personalized information and advertisements to users. This method is also used in the industrial field. The Internet of things and artificial intelligence can label both production processes and consumer users to meet the most refined needs. For example, 3D printing, in essence, can flexibly change processes according to various needs without relying on physical models and can print new products solely by computer modeling.

Therefore, the connotation of industrial automation has been changed; it no longer automatically produces for fixed demand, but it automatically regulates production, circulation, and distribution according to changes in demand. Fei-Yue Wang, the first Chinese recipient of the Norbert Wiener Award in the field of cybernetics and deputy director of the Institute of Automation of Chinese Academy of Sciences, said, "Industrial automation will shift to knowledge automation." The new production process will have the following characteristics: automated tracking of human life diversity, automated acquisition of knowledge, self-use and evolution of tools, automatic optimization of social management, automatic adjustment of production processes based on knowledge, and production of new knowledge. A new cycle of reciprocation will be formed, and the industrial economic paradigm will be innovated. This process will challenge all the past production methods, from production layout, design process, and niche construction to bureaucratic corporate power structure. "Made in China" will become "intelligently made in China."

Toward the Internet of Things and Refined Production

The basis of knowledgeable, automated, and refined production methods lies in the Internet of things.

In July 2016, Softbank Group spent £24.3 billion to acquire microchip giant ARM Holdings. Softbank President Sun Zhengyi believes that the Internet of things will lead toward the next round of technological explosion. In 2018, the number of IoT devices will exceed that of mobile devices; it is estimated that in 2021 there will be 1.8 billion PCs, 8.6 billion mobile devices, and 15.7 billion IoT devices in the world. In 2035, the amount of data will be increased by 2,400 times, from 1EB (exabyte) to 2.3ZB (zettabytes). In the next twenty years, the number of IoT devices will exceed 1 trillion. Sun Zhengyi pointed out, "The relationship between the Internet of things and artificial intelligence is just like the eye and the brain cooperating for biological evolvement. The explosion of IoT is about to come."

The IoT connects people and everything to the same network, allowing people to interact with the machine at any time. All the behaviors and reactions that can be digitized may cause changes to machines and production lines. Various types of data gather in the cloud, and a large number of calculations are completed through the cloud server before being sent back to people through products and services. This cycle thus links people's diversity with the material world, spawning each other and coevolving.

Ray Kurzweil, author of *The Singularity Is Near: When Humans Transcend Biology*, believes that human beings are in the era of industrial revolution, which is led by the Internet of things. The 3D printing technology is one of the main driving forces for the innovative industrial revolution before 2020. By 2030, our clothes may be designed as an open source and directly downloaded from the cloud for free. The future of 3D printing technology is unclear, but Kurzweil's words echo the idea of the new industrial paradigm for diversified needs. In addition, from food to music, everything can be refined through the Internet of things, with cloud computing as its internal logic.

Kevin Ashton's views are more profound. He states the following:

We have to distinguish between smart wearables in popular terms and the Internet of things. The bikini that can detect sun protection index and smart cups that can sense thirst are not Internet of things. The Internet of things is not a terminal device, but a set of machine systems that can learn independently and make decisions on their own.

The advantage of the IoT is that computers have a variety of sensors that can collect data by themselves. Just as our smartphones have GPS or BeiDou Navigation Satellite System, maps, and sensors of distance, direction, gravity, inertia, and even heart rate, all the information can be collected for processing. Fast-growing RFID chips can provide data without power consumption. The output of RFID has already exceeded that of mobile phones, giving each mobile phone and each device a unique code, or "name." Through this nonconsumable RFID system, hotels or parking, reservation, and meal payment can be completed, which is also a component of the IoT. In the future, the power consumption of IoT devices will be extremely small, and mobile phones will even charge themselves in the wind.

The human brain will connect directly to the cloud and become part of the IoT. This will promote the reverse-engineering analysis of the brain's way of thinking and not only deepen the understanding of the human brain but also improve machine intelligence in reverse, and understand the diversity of people more deeply.

The industrial robots of the past, which mainly rely on industrial automation, will undergo morphological changes in the IoT era. Industrial robots will evolve from being hardware-oriented to software-oriented, and then the virtual state of the cloud will be combined with the physical entity. Robotic software and the IoT are the same but are developing in two directions. Machine intelligence will be invisible and ubiquitous, and, combining with cloud computing, will become a huge social robot. This is the ideal pursuit of artificial intelligence and the IoT.

The Internet of things system is fundamentally different from the traditional machine system. In terms of operational logic, it is the sublimation from the mechanical Newton machine to the intelligent Merton machine. The Newton machine runs according to the law of causality, while the Merton machine learns the law autonomously according to the relevant thinking. The Newton machine follows the logic of "big law, small data," whereas the Merton machine follows "small law, big data," works more like the human brain, and copes with the rapid changes of the world.

The independent collection of data by the IoT is only a basic task, and the higher realm is self-determination. Humans provide algorithms and training models to equip the machine with sensors that allow it to judge and communicate in its own contexts and make decisions. Decisions affect the world, which leads to new data, thus forming a cycle.

For example, Baidu Cloud Computing has already supported multiple third-party smart-lighting projects. Today's urban outdoor lighting system is bright, but most of the time it shines the empty roads and the sky. The IoT lighting system can collect data at an early stage, automatically learn the lighting rules, and then optimize the system independently. For example, some street lights are autonomously dimmed and turned off during the low-traffic period. The system monitors the status of the street lights in real time, predicts the life of the equipment through machine learning, and makes an accurate statistic for the parts that need to be updated or replaced, thus reducing the inventory and saving maintenance costs by up to 40 percent. Ultimately, power consumption is reduced up to 40 percent compared to traditional standards, and the equipment life is extended. In this way, the past lights have become the breathing rhythm of a living body.

Economic results are only part of the system's benefits. The aforementioned lighting system integrates intelligent lighting control, environmental sensing, wireless city and security functions, and opens up a large number of APIs for more applications. Because the computing power and storage capacity of the Baidu cloud platform can be expanded, the system has obtained great flexibility. Instant analysis of hot data and big-data mining of cold data can be completed. Such a system is not just for lighting but also serves as part of a smart city. It promotes machine intelligence and the city's brain by collecting data, learning on its own, and operating by itself.

The IoT will cover all living spaces of human beings. In a sense, we can also understand smart agriculture as IoT. In some advanced demonstration farms, each plant is equipped with sensors. For example, the system will water and fertilize according to the feedback information of each plant, which will greatly improve efficiency and save resources. The sensor of each plant is connected to the system, and the large-area crop sensing information is aggregated into the cloud, so the agricultural methods beyond the farmer's experience can be calculated to realize the agricultural revolution.

People may understand unmanned vehicles as innovations in the transportation field. But the unmanned vehicle system goes beyond the traffic field and becomes a carrier of the city and intercity of the IoT. An unmanned car is not an isolated machine, but a large, autonomous system that connects all other urban systems like blood vessels and nerves. The unmanned vehicle itself is an aggregator of artificial-intelligence technology. It integrates visual recognition, speech recognition, autonomous decision-making, and mechanical control. It is a motion data collector and processor. On this basis, the unmanned vehicle network binds people, cars, and the environment, linking personal goals to overall management. Once the unmanned vehicles are put into large-scale use, they will drive the development of technology and IoT. Every component sensor and passenger sensor on an unmanned vehicle is associated with a manufacturer, consumer, manager, or even a third party. Imagine that in the future the land transportation with mainly unmanned vehicles will open up information with aviation and maritime traffic, and that will be a huge Internet of things spreading from up in the sky to down to the ground.

If the refined production of the C-end (consumer) can respond to the diverse requirements of people, then the B-side (business) of the IoT can refine the overall demand of the society. Artificial intelligence and the IoT work together to enrich production, increase efficiency, and reduce external costs. For example, smart agriculture significantly saves water and fertilizer, and the unmanned vehicle system will reduce vehicle accidents, guarantee basic travel, and reduce pollution.

Smart energy, smart transportation, and smart production will bring unprecedented changes in the world.

Calling for Intelligent Government and Intelligent Society

After experiencing the industrial revolution, two world wars, and countless changes and turbulence, human society has increasingly recognized that individual freedom, stability, and development are inseparable from the efficiency and justice of the government and society. With the development of the economy and society, organizations are becoming more and more complex, which requires new ways of governance. The role of government and social organizations is particularly important.

Modern society adopts laws to maintain and regulate social relations. However, the development of technology, especially artificial intelligence, has increased the value of algorithms. Various automated management tools have delicately regulated human interaction, consumption, transportation, and finance through algorithms. In future society, the law may be integrated into the algorithm.

Large-scale data governance originated from government information management since the twentieth century. For example, in 1929, President Herbert Hoover advocated a nonelectronic crime information-recording system in the United States. During the 1960s, the United States began to establish a nationwide unified crime-information system. The use of the data went beyond the criminal record check, which greatly helped the labor-market selection and welfare-plan implementation, thus becoming a cornerstone of governance.

In the future, with the development of artificial-intelligence technology, the governance model and the rule-of-law structure may undergo major changes. Zheng Ge, a professor of law at Shanghai Jiao Tong University, said the following:

> The [current] law generally assumes that responsibility stems from fault. The fault damages the rights and interests that laws determine to protect, and causes harmful consequences, so there comes legal remedy. Based on this principle, the law always lags behind the damage. Only when the subjective fault promotes the specific behavior and the behavior causes the actual damage can the law intervene, and the purpose of the intervention is to restore the previous state. The emergence and popularization of the Internet has changed people's way of communication and interaction, and the development of big data technology has brought the potential of the Internet to a new level. The combination of big data technology and cognitive science and artificial intelligence makes behaviorism likely to become obsolete. Predictive and guided data

analysis can change the application field of the law by personalizing the identification, analysis and intervention of "implanting" intentions and behavioral motives.[11]

This is a vision of the future shown in the American film *Minority Report*—government agencies may predict crime through data and stop it in advance, rather than chasing the criminal afterward. We can think that many management methods of the future government will probably need to change from chasing management to predictive management.

Reports on the development of artificial intelligence in both the United Kingdom and the United States have already stated such help or challenges that artificial intelligence brings to governance.

The UK government report states that it is already implementing data science technologies, such as machine learning, which provides deep insight with a range of data, from digital-service feedback to satellite-imagery analysis. For example, the government can do the following:

- Improve the efficiency of existing services (such as health, social security, and emergency services) by predicting demand and more accurate customized services, so resources are allocated to the maximum extent possible.
- Make more reference data accessible to government officials so they can make decisions and reduce the probability of fraud and error.
- Make decisions more transparent (through digital records behind the acquisition process or data visualization to support decision-making).
- Help government departments better understand the people they serve and ensure that everyone is provided with appropriate support and opportunities.

The White House report suggests that artificial intelligence can be used to improve the criminal justice system. The government should take steps to promote the full application of law-enforcement data and public data, so that the algorithm system can get better to help people reduce prejudice in all

---

11 Zheng Ge, "Between Encouraging Innovation and Protecting Human Rights—How Does the Law Respond to Challenges from Technical Improvement of Big Data?" *Exploration and Free Views*, no. 7 (2016).

aspects, such as crime reporting, public order, establishment of bail, sentencing, and parole decisions, and to make efficient and fair decisions.

Scientific research institutions in the United States are trying to use artificial intelligence to solve economic and social problems, such as using data mining and artificial intelligence to improve unemployment, lower dropouts, and help homeless people. Researchers at Stanford University are using satellite intelligence to analyze satellite imagery and provide guidance for assistance to poor areas.

The UK report suggests that government agencies can use artificial intelligence to alert people about fire hazards that may occur in cities. This has already been achieved in the United States.

There are approximately one million buildings in New York City and an average of three thousand buildings have had serious fires every year. Since it happens every year, the fire department wondered if it could predict the fires.

Chinese data scientist Jeff Chen was a data analyst at the New York City Fire Department. He believed that each building has unique properties that can be analyzed to find out which are prone to fire. For example, houses of low-income families are more prone to fires, and the high population density makes the fires more harmful. Other fire-fighting factors include the age of the buildings and the circuitry, equipment of fire-fighting facilities (the number and location of fire hydrants), and elevators. Chen said that vacant or unsecured buildings are twice as likely to catch fire as other buildings. These points seem obvious, but they are not so easy to interpret through data analysis.

Chen led the development of a predictive fire-risk software engine, analyzing the various data about buildings and households through mathematical statistics, supplemented by machine-learning techniques. He drove the engine through municipal data and predicted the fire risk of different buildings. The system was deployed during 2013 and integrated nearly 7,500 real-time risk factors. The New York Fire Department used the system to calculate the fire hazard for 330,000 buildings that could be inspected (the fire station does not inspect single-family or double-family homes).

Prior to this, fire inspections were random. Now, when firefighters conduct weekly routine inspections, the system generates a list of buildings sorted by risk factor, which helps firefighters prioritize inspections of buildings prone to fire, saving manpower and increasing efficiency. In addition, the data-analysis system also was involved in garbage disposal and solving social security problems.

At the government level, smart governance projects are more prominently used for security work. In addition to being everyone's object of criticism, the CIA has invested in a number of unicorn companies in the data field, including Palantir Technologies, which Peter Thiel invested in, as well as Dataminr, TransVoyant, Geofeedia, and PathAR.

The principles of these companies' products are similar; they are automatically collecting social data through social media, maps, sensors, and other channels and integrating various types of transportation and financial disclosure, connecting the separate databases, and providing various pivoting solutions without user coding.

Palantir Gotham, which belongs to Palantir Technologies, is mainly used for counterterrorism, and this derives from PayPal's fight against fraud due to their similarity in data collection and matching. It develops big-data tools to discover and suspend suspicious accounts by matching data such as past transactions and current fund transfers. Then, Palantir Gotham thought this technology could serve the government as well. Intelligence agencies such as the CIA, FBI (Federal Bureau of Investigation), and DIA (Defense Intelligence Agency) and the army, navy, air force, and police have thousands of databases, including financial data, DNA samples, voice material, video clips, and world maps, but it is quite cumbersome to establish links between these databases and find valuable information. Palantir's founding team believed that if it built a data-analysis library and integrated separate data for search and analysis to improve data-analysis efficiency, then it could sell the technology to the government.

Geofeedia can quickly dig up all kinds of information about a place where the news is happening. TransVoyant collects data intelligence through a variety of traffic sensors and map technologies and integrates local news, social-media information, weather reports, satellite imagery, travel warnings, criminal-activity information, etc., to provide forecasts for the government, help with asset management, and make data decisions in real time.

The intelligence service aspect of technology companies carries over to their business side. Selecting target customers, saving time and adding benefit, getting information before the public, and predicting the future, which is in line with the high-frequency trading in financial speculation, combat decisions, and stock markets, are examples of business applications. This is the jungle model of the free market, rather than the governance model of coordinating the overall situation of society, which must be considered.

China has relatively favorable conditions. Qi Lu believes that "China is gradually becoming a big innovation country in emerging industries and there are many opportunities for innovation, while the United States is relatively conservative. In the era of great change, China has a large space for innovation and provides a better environment for innovation." China's smart city will become a pioneer in intelligent governance, due to China's strong leadership role in government departments, with smart maps, security cameras, and data-management systems covering urban and rural areas.

6

RUSHING INTO NO-MAN'S LAND OF AI: THE ROAD TO THE UNMANNED VEHICLE

Artificial intelligence can only reflect the revolutionary significance of technology if it is closely linked to the fate of mankind and confronts complex situations. Milan Kundera said, "The heavier the burden, the closer our lives come to the earth, the more real and truthful they become." Unmanned vehicles are the chief artificial-intelligence project that clings to the earth. "Difficult" and "subversive" are two adjectives that the concept must overcome. Unmanned vehicles were born before artificial intelligence, but they face more hurdles before they can exist in a meaningful way. The rut left behind them will be the social order since the birth of transportation.

An unmanned vehicle, also called an autonomous vehicle, is a car that senses the environment and navigates without human intervention. When the first unmanned vehicle was on the road in the United States in 1925, the inventors had not reached a consensus on whether it needed to be smart. But since then, the image of unmanned vehicles has appeared in different kinds of science fiction and movies. This dream is so specific, but after several generations it has not come true yet. Fortunately, the frustration of every dreamer makes the road of unmanned vehicles easier; this is a road to intelligent evolution and a road of change for human society.

In the future, unmanned vehicles will no longer be just transport tools; they will also be a third space outside the home and office, which is mobile, safe, and comfortable. People will move in space at the lowest cost. Traffic jams, pollution, and difficult parking caused by inefficiency will be greatly alleviated. Driving behaviors that endanger the safety of others, such as drunk driving, driving through red lights, and speeding will no longer exist. The unmanned vehicle system will likely become the blood vessel of the global Internet of things system, and even the social status and symbolic meaning of cars as a mode of transportation will be completely changed.

According to the World Health Organization, in 2015, more than 260,000 people were killed in traffic accidents every year in China, which ranks first in the world—about 90 percent are caused by human error. Unmanned vehicles are expected to reduce the number of accidents down to 1 percent of the current level.

There are also huge gains. In a report, Morgan Stanley pointed out that after the development of unmanned vehicles, the United States can earn $1.3 trillion in revenue, which is equivalent to 8 percent of US GDP. More than $100 billion will come from fuel savings, more than $200 billion from reduced traffic jams, more than $500 billion from medical and insurance costs by reduced traffic accidents, and more than $400 billion from higher work efficiency.

But there's more. Once smart tools connect cars and cars, people and people, and people and society, they will reshape the rules of the material world.

Forging Ahead on Rough Roads

Back in August 1925, no one was in the driver's seat of the car named American Wonder, and the steering wheel, clutches, brakes, and other components all "act[ed] according to circumstances." Engineer Francis P. Houdina sat in another car behind it and used the radio waves to control it. The two cars traveled through New York's crowded traffic, from Broadway to Fifth Avenue. This experiment, which can almost be regarded as a super remote control, is still not widely recognized by the industry today.

In 1939, skyscrapers began to appear in the United States. People who gradually regained their confidence after the Great Depression had a vision for the future. At that year's World's Fair in New York, people lined up outside the GM's Futurama (Future World) pavilion to see what the "future" would look like. Designer Norman Bel Geddes showed his imaginary car to the people—it was radio-controlled and powered by an electromagnetic field embedded in the road.

He further explained in his book *Magic Motorways,* published in 1940, that humans should be separated from driving. American highways will be equipped with something like a train track to provide an automatic driving system for cars. When the car will travel on the highway, it will move forward following a certain trajectory and program and resume human driving when off the highway. For this idea, the start date he gave was 1960.

During the 1950s, researchers began experimenting with these ideas. Perhaps the experiment made people become aware of the difficulties, and the predictions were then postponed—they said unmanned driving would start in 1975.

After that, attempts were made to use the wires laid on the ground to navigate the car, and the technology exploration of unmanned vehicles was carried out everywhere. The British installed guide wires that were composed of permanent magnets. They thought the wires would make the control more precise and the speed faster. But, the "ground orbital" experienced numerous setbacks, and almost everyone came to believe that the orbital guidance of unmanned driving did not have practical application value. The trials were already at the ceiling of the technical expertise at that time. However, this technology is implemented in some food delivery robots.

In 1956, GM created the Firebird II concept car, the first with an automatic navigation system. The streamlined body made of titanium looked like a rocket, as if the idea of that car came directly from science fiction movies. When afterward the Firebird came out in the third generation, the slogan was "Want to sit and relax? Well, set the speed, then adjust to automatic navigation. Release the handle. Firebird III will nail it."

GM invited the BBC for live broadcast of the unmanned driving test on the highway. However, the unmanned vehicle still traveled by receiving an electronic pulse signal sent through the embedded cable, failing to get rid of the "ground orbit" idea.

At that point, both the remote control in the hands of humans and the not-so-feasible preapplied cable were far from the free and smooth unmanned driving experience in human imagination. In the 1960s, Robert L. Cosgriff, a driverless-car project leader at Ohio State University, was convinced that electronic navigation devices would be embedded in public roads within fifteen years. Laboratories of countries all over the world worked on the navigation devices for many years.

Twilight Is Not Far Away

At that time, the basic technology used in today's unmanned vehicles had already emerged in major research institutions. But those technologies were scattered all over the place, and no one thought about combining them.

In 1966, intelligent navigation first appeared in Stanford University Research Institute. Shakey the robot, developed by the SRI Artificial Intelligence Research Center, moved on a wheel structure. It could take hours to complete the simple action of switching lights, but the built-in sensors and software systems pioneered the function of automatic navigation.

In 1977, the Tsukuba Research Laboratories in Japan developed the first self-driving car that was based on a camera to detect marks or navigation information ahead. The car was equipped with two cameras and could reach up to 30 kilometers per hour (a little over 18 miles per hour) with the aid of overhead tracks. People began to think about the future of unmanned vehicles from a visual perspective. Navigation and vision together put an end to the ground orbit.

A bit earlier, in 1973, the GPS system began to develop. DARPA (the Defense Advanced Research Projects Agency of the United States) launched the ALV (Autonomous Land Vehicle) program in 1984, with the purpose of detecting terrain through a camera to calculate navigation and driving routes by a computer system. The robot used radar to identify roads, navigated by GPS, and used miniaturized short-wave radar to detect sudden obstacles in front before automatically braking. It is not difficult to see the maturity of the unmanned vehicle research, but, unfortunately, the research of the ALV project was forced to terminate after five years because Congress cut the funding due to its limited results. This termination delayed the birth of the unmanned vehicle for several years.

Germany also invested in unmanned vehicles for the military. German military research institutes had been working with Mercedes-Benz since 1987 to develop unmanned vehicles. The technology, which uses cameras and computer image-processing systems to identify roads, is even more mature than DARPA's ALV project. But the study did not achieve much either.

Today, we know that to actualize the dream of the unmanned car we must overcome countless technical gaps of computing speed, big data, and deep learning.

THE ROAD TO THE UNMANNED VEHICLE

From 1993 to 1994, the team of Professor Ernst Dickmanns from the Bundeswehr University of Munich, Germany, modified a Mercedes-Benz S500 sedan and equipped it with a camera and various other sensors to monitor the environment and changes around the road in real time. This was the most successful "dynamic vision" experiment of the era. The Mercedes-Benz S500 automatically drove more than 1,000 kilometers (more than 620 miles) in normal traffic conditions.

Almost at the same time, Carnegie Mellon University, which began investing in unmanned vehicles in 1984, pioneered the use of neural networks to guide autonomous vehicles in 1989, even though the server for the refurbished military ambulance in Pittsburgh was as big as a refrigerator and only about one-tenth of the computing power of the Apple Watch. But in principle, this technology is in line with today's unmanned vehicle-control strategy.

The university's Navlab project developed the fifth-generation unmanned vehicle in 1995. A 1990 Pontiac Trans Sport had been modified to include a laptop, windshield camera, GPS receiver, and some other ancillary equipment. It successfully completed an unmanned journey from Pittsburgh to Los Angeles. In the present sense, it can be counted as semi-automatic driving, though the research results are meaningful for the current unmanned technology.

Since the 1980s, China has researched intelligent mobile devices, and the initial projects started with the military. In 1980, China established a remote-controlled antinuclear reconnaissance vehicle project. The Harbin Institute of Technology, Shenyang Institute of Automation, and National University of Defense Technology participated in the research and manufacture of the project. In the early 1990s, China also developed the first driverless car in its true sense.

With the support of the 863 Program, many universities and institutions began to study unmanned vehicles. After the "Eighth Five-Year Plan," "Ninth Five-Year Plan," and "Tenth Five-Year Plan," three generations of ATB (Autonomous Test Bed) series of unmanned vehicles were developed by five organizations, including Beijing Institute of Technology and National University of Defense Technology. ATB-2's straight driving speed reached up to 21 meters per second (almost 47 miles per hour), and ATB-3 had further advances in environmental awareness and trajectory tracking.

The Red Flag CA7460 driverless car, developed by the National University of Defense Technology and FAW Group, can automatically change lanes on the road according to the situation of the vehicles ahead, and the driving speed can reach up to 47 m/s (about 105 mph). The THMR-V unmanned vehicle developed by Tsinghua University has a highest driving speed of about 42 m/s (94 mph), and can also choose two driving modes—highway and urban—according to different driving scenarios. Springrobot, developed by Xi'an Jiaotong University, is also one of the famous unmanned vehicle platforms in China, with a highway line-detection capability and pedestrian-detection capability. The National Natural Science Foundation of China launched the Cognitive Computing Audiovisual Information major research project in 2008. In 2009, the first China Intelligent Vehicles Future Challenge was held in Xi'an, and in the following years, dozens of driverless cars from famous universities and research institutions participated in the competition.

Such events started earlier abroad. Driverless cars showed signs of rapid development in the 2004–2007 DARPA series of competitions.

In 2003, the United States started the Iraq War, which ignited the fuse for unmanned-vehicle technology. Transporting ammunition through the desert has become normal for the US military, but even with armored vehicles and helicopter protection, the heavy fleet is still frequently attacked. Furthermore, US military personnel often encounter roadside bombs or landmine attacks, which cause heavy casualties. As a result, DARPA, which once spawned the Internet, restarted research on autonomous driving technology that had been on hold for more than a decade. Ten years later, at least one-third of military vehicles can drive automatically.

In 2004, the US Congress allocated $1 million, and then added $2 million, for an unmanned driving race held by DARPA and called for participants. Although the prize, compared with the cost of software and hardware for unmanned vehicles, was not much, the race still attracted many contestants, including business enterprises, research institutions, university educational institutions, and even individuals.

As required by the war, the first unmanned car challenge was held in the desert. The route, from Barstow, California, to Primm, Nevada, was about 230 kilometers long (143 miles), mostly consisting of desert, which made driving difficult. The game was required to be completed within ten hours, only relying on GPS to guide driving, and sensors or cameras to bypass natural obstacles. The military decreed that only one would be champion.

We don't know if the military underestimated the difficulty of unmanned-vehicle technology or overestimated the capabilities of the participants. No team arrived at the finish line. Even the first-place vehicle in the competition, the Sandstorm of Carnegie Mellon University's Red team, only travelled 11 kilometers before it turned toward a wrong direction and got stuck in a ditch. Most of the participating vehicles failed at the beginning of the race, with malfunctions of brake locks, axle breaks, rollovers, and satellite-receiver failures.

The results were very frustrating; people's expectations dropped sharply, and pessimists said that unmanned vehicles that could travel in the desert might never be made. However, the competition did not stop there.

In 2005, the second unmanned-car challenge was again held in the desert, near the junction of California and Nevada. The course was 212 kilometers (132 miles), with an increased difficulty level—three narrow tunnels and more than 100 sharp turns. The most difficult section was a narrow bend, with a ditch of more than 60 meters (almost 200 feet) deep on one side and a cliff on the other. However, this time the vast majority of participating cars exceeded the previous record of 11 kilometers. Although many participating vehicles withdrew by the midway point, five of them finished the whole course.

The top four came from Stanford University, Carnegie Mellon University (with two vehicles), and a private US company. The champion vehicle in this competition was equipped with cameras, a laser range finder, radar remote ranging, a GPS, and many other essential equipment for today's unmanned vehicles.

In 2007, DARPA moved the location to the city and began to transfer the focus from military to civilian. Complex environments, including traffic lights and cars, in the George Air Force Base are like the environment in which future unmanned vehicles will actually be used. At the end of the competition, Carnegie Mellon University, Stanford University, and Virginia Tech won the top three. Unmanned vehicle research seemed to have stabilized, but the DARPA event ended abruptly. For various reasons, the US military has not used unmanned-vehicle technology for logistics. However, once the engine of the technology starts, no one can stop it.

The champion of the DARPA Unmanned Car Challenge in 2007 was the Boss from Carnegie Mellon University. The size and complexity of the equipment on the roof and in the front of the car far exceeds that of all types of unmanned cars today. DARPA has unearthed the potential of unmanned-vehicle researchers and has also guided the basic technology.

The participating teams generally used cameras, radar, and laser equipment to detect surrounding terrain and obstacles. The results were integrated with GPS and a sensor information system to make operational decisions, such as acceleration, deceleration, turning, etc. After more than ten years, the successors to this technology have done nothing more than refine the basic paradigm.

This competition promoted an unmanned-vehicle system consisting of inventors, engineers, programmers, and developers, which contributed to the rise of the unmanned-vehicle technology-investment boom. Google, Apple, Tesla, Uber, and Baidu successively have all announced their plans to develop unmanned vehicles. Google headhunted not only Sebastian Thrun, the leader of the Stanford team, but also the entire research group of unmanned vehicles at Stanford. Many students and colleagues of Red Whittaker, the core figure of the 2007 championship team, have become cornerstones of the US unmanned car industry at Carnegie Mellon University.

With the progress of Internet companies in the unmanned car industry, even conservative traditional car manufacturers joined the fray.

Although the high cost of experimentation restricted mass production and commercialization, 2007 is still worth remembering, not only because it opened a new chapter in unmanned vehicles but also because deep-learning research gained new life. Afterward, the related fields have also shown a breakthrough: the big-data revolution, the rise of cloud computing, the wave of mobile Internet, and the diversification of data collection channels.

More changes have continued to free the unmanned car from the shackles of the traditional automotive industry and get it out from university labs.

Over the years, internal combustion engines, gearboxes, and complex production processes have created a high barrier to the traditional automotive industry. However, the maturity of new energy vehicles has eased the situation. Electric cars no longer require the most technically difficult engine, transmission, and clutch. This not only gives some technology companies the opportunity to delve into this field but also makes it possible for the automobile manufacturing industry in some lower-end countries (such as China), long trapped in the core technology gap, to have the opportunity to be competitive.

Long-Established Car Companies at the Foot of the Mountain

In 2013, when smart driving was ascending, the US National Highway Traffic Safety Administration (NHTSA) officially issued a classification. According to the given definition, the development of smart cars can be divided into four levels:

- Level one (L1) is the "advanced assisted driving system," which is characterized by providing drivers with collision warning, emergency braking, blind-spot monitoring, and weak-vision intensification for night driving.
- Level two (L2) is "automatic driving in a specific environment," close to GM's vision that the vehicles would be driven automatically in a relatively regular environment, such as highways or traffic jams.
- Level three (L3) gives some outlines of the expectation of unmanned driving. L3 is "automatic driving in multiple environments." The L3 vehicles would adapt to all road conditions, but in special cases human drivers would need to take over.
- Level four (L4), the "automatic driving" level, designates smart cars that can actually drive by themselves. The difference between L3 and L4 is that the L4 smart car will not have the steering wheel and brakes.

The classification can be regarded as a reshuffle in the automotive industry. In the world of unmanned vehicles, traditional car companies, Internet companies, and taxi-industry giants are three main powers. The interest of traditional car manufacturers lies in L1 and L2, while L3 and L4 attract giants like Google, Baidu, Uber, and Tesla.

Today, the decisive factor is no longer capital or history but the intrinsic nature of each company. Between L2 and L3 is a technological gap difficult for long-established car companies to cross. The growth of technology, data, and talented workers of Internet companies, including Baidu, has given the technological companies a lead that traditional car companies can't overcome in a short period. But even with similar devices, such as cameras, sensors, deep learning, radar, etc., the varying quality of software databases mean different levels of unmanned driving effects.

The BMW I3 is equipped with four IBEOs (laser sensors) that are used for emergency braking and avoiding column obstacles. Unfortunately, the usage seems stuck in a multilevel parking-lot environment. Volvo's XC90 is equipped with City Safety, which adds a new out-of-road protection system and an automatic brake system at crossroads to help drivers avoid common rear-end collisions.

In early 2015, Mercedes-Benz launched the F015 concept driverless car, which was filled with various displays and rotatable seats, making it a mobile entertainment center.

However, the Mercedes-Benz F015 still cannot get rid of the shackles. The Intelligent Drive system is still limited to be designed to prevent collisions, stay in lanes, apply autobrakes, and use auto-following functions in traffic jams, which are not enough for modern driverless application.

Under the leadership of CEO Mary Barra, in 2015, GM acquired Cruise Automation and invested $500 million in the taxi-service company Lyft to launch GM's first long-life, self-driving all-electric car, Bolt. Electric cars, autonomous driving, and sharing economy—every part of that is fashionable.

But the industry was perplexed with moves of General Motors. Pundits thought that once a traffic accident occurs with an unmanned car, it will put car enterprises in a very disadvantageous position. In addition, a taxi service with unmanned vehicles is likely to significantly reduce private car ownership. Car companies would then lose prestige, becoming the most basic hardware providers. So Barra's ambition was ridiculed as just an attempt to calm anxiety over the coming wave of unmanned cars.

German car manufacturer Daimler reached an agreement with Nevada in 2015; the state licensed two driverless trucks for use on public roads. Daimler was the first company to send unmanned trucks on the road. In April 2016, Daimler arranged for its three self-driving truck fleets to drive from Stuttgart, Germany, to Rotterdam, the Netherlands. However, the route was limited to highways, and there was still a driver in the car to supervise.

Marginal Advance of Smart Enterprises

Before being acquired by GM, Lyft had partnered with Chevrolet to develop unmanned vehicles. Even BlackBerry turned to unmanned driving after abandoning the mobile-phone industry. These Internet companies just stepped into the auto industry without the hesitation of auto-industry giants and are eager to invest in the research. In 2016, the more radical Tesla has sent more than one hundred thousand semi-autonomous vehicles on the road, which can overtake and avoid obstacles and automatically enter parking spaces.

A young technology naturally can cause problems. In January 2016, the first accident of Tesla's self-driving car happened in China. Although the investigation report is unavailable, we do know that this car, with an automatic driving system, did not try to slow down when a car swept in front, resulting in rear-end collision. In May 2016, a Tesla self-driving car had an accident in the United States. Under strong sunlight, the autopilot system failed to identify a white container truck that had crossed in front of it. It did not brake in time, although the Tesla driver violated the operation specification by having both hands off the steering wheel. This accident killed the Tesla driver. In August 2016, two drivers in Texas and Beijing, respectively, crashed into the high-speed guardrail or smashed with a car on the side. Post hoc analysis indicates a sensor-system misdetection.

In the accident of the car hitting the white container truck, the truck in a high-brightness sky background was mistaken for white clouds by the visual-recognition system because of the multiple lanes and the frame being above the ground. In the Beijing accident, the roadside stationary vehicle was recognized as a fence by the identification system, and the system miscalculated the safety distance. This shows an insufficient training of the identification system.

Uber began testing unmanned vehicles without obtaining a California government road-test license. But the experiment turned into a farce because the vehicle almost crashed after running a red light.

In the unmanned-driving industry, the car updates road data in real time, with embedded artificial-intelligence solutions included. The goal of the updates is to enable automatic cars to achieve road awareness without being connected to the Internet.

Most of the attempts discussed here aim at L3 unmanned vehicles. But, for Baidu and Google, the real ambition lies in L4. The L4 level has an extremely high entry barrier, and artificial intelligence and high-precision map information are indispensable. Therefore, Google and Baidu have a distinct advantage over the competition.

In 2009, Google began research and development of its own driverless-car project through the DARPA's support. In December 2014, Google released a completely self-designed driverless car. The exterior structure was completely different from the traditional car, without a steering wheel or brake. By 2015, the prototype was ready to go on the road; passengers could wait for the vehicle to start as soon as they were sitting in the car. By October 2016, this white car, which many people described as "cute," had carried out more than 3.2 million kilometers (just shy of 2.0 million miles) of public road testing, equivalent to 300 years of driving experience for human drivers.

Probably, Baidu can understand Google's ambition the best. Almost concurrently with Google's car, Baidu's unmanned vehicles had taken off. During the robust development of China's mobile-Internet industry, Baidu had been immersed in the development of unmanned vehicles, without publicizing it to the outside world—to the public, Baidu's attention was on security and learning-outcome evolution.

Like Google, Baidu wants to achieve the completely driverless vehicle, and it has chosen the most difficult technical road: high-precision map and sensors. That method is several orders of magnitude more difficult than that of all the aforementioned L3 auto-driving cars. Baidu Brain has become instrumental in Baidu's plan.

Baidu initially kept the project highly confidential. Until December 2015, the Baidu driverless car completed the automatic-driving test on Beijing highways, and then the outside world gradually learned what Baidu was doing. The Baidu driverless car became even more familiar to the public because of a ticket: on July 5, 2017, as a part of the Baidu AI Developer Conference, I took the driverless car on the Fifth Ring Road of Beijing and conducted a live broadcast. On the same day, we received a ticket issued by the Beijing Municipal Commission of Transportation. This was the first ticket given to a driverless car in the world. It made the Chinese people realize that the era of autonomous driving is really coming.

During the test on Fifth Ring Road in 2015, Baidu's driverless car achieved automatic driving in mixed city road conditions—including jug handles, and highways—for the very first time. The Baidu driverless car departed from the Baidu Building in Zhongguancun Software Park, entered the G7 Jingxin Expressway, arrived at the Olympic Forest Park via Fifth Ring Road, and then doubled back. It operated under automatic driving for the whole process and utilized multiple driving actions such as following, decelerating, changing lanes, overtaking, going in and out of ramps and turning around, and entering and leaving the highway. The test speed was up to 100 km/h (or 62 mph).

The roof of the Baidu driverless car was equipped with a sixty-four-line laser radar on the roof, which performs a panoramic scan of the road conditions within a 60-meter radius of the car through continuous circumferential rotation. Three laser radars are installed at the front and end of the vehicle to scan adjacent areas to compensate for the blind spot of the roof radar. This series of high-resolution laser radars has a thorough overview of the overall road conditions, which helps to achieve autonomous driving in complex traffic conditions, especially for special scenes such as traffic congestion, narrow roads, residential areas, and parking lots, to achieve following, lane changing, and intersection crossing.

But the lasers are not enough; there are two visual-recognition cameras in front of the roof, so that this pair of "eyes" can accurately identify traffic lights, road traffic lines, and traffic signs, making up for decline of laser radar in special weather conditions such as rain, snow, fog, etc., and improving the recognition of road signs and signals.

Because of the environmental awareness, behavior prediction, high-precision maps, and high-precision positioning, an unmanned vehicle gathers the best technology in many fields. Baidu can advance in so many different directions by relying on long-term accumulation in the field of artificial intelligence and deep learning. Of course, looking at the history of driverless technology, we can also say that Baidu's driverless technology stands on the shoulders of giants. Each development in technology is the optimal solution to the repeated failures of other researchers.

Today, Baidu's driverless cars are carrying out a large number of road tests in China and the United States. According to Baidu's timetable, during 2018, the unmanned microcirculation bus (or "community bus") will be mass-produced with King Long and auto-driving models will be made with JAC Motors and BAIC Motor in 2019, and with Chery Automobile Company in 2020.

The research is not the only mark of progress for Baidu; achievements of Chinese manufacturing stand out as well. From the very beginning, Baidu undertook the responsibility of more than just research.

At the World Internet Conference in 2015, the driverless car displayed by Baidu attracted the attention of President Xi Jinping. President Xi stayed at Baidu's booth for more than 10 minutes, three times longer than the scheduled time. Li Deyi, an academician at the Chinese Academy of Engineering, once analyzed the unmanned-vehicle industry. He believes that "the wheeled robot represented by the driverless car will become the first business card of China's smart manufacturing by 2025." Unmanned vehicles are not only vehicles but also the foundation of the Internet of vehicles and the Internet of things. The equipment systems will directly drive the rapid development of radar, sensors, navigation systems, and other industries in the process of commercialization. The strategic value of the unmanned vehicle has far exceeded that of just the vehicle.

China has a rich traffic environment, a large population, and a huge market. Unmanned vehicles will not only bring changes to the entire transportation model but also bring new things such as smart subways and intelligent public systems in China. Smart transportation is a systematic, revolutionary, and subversive change. With the rapid development of technology, perhaps people's thinking should be changed.

Qi Lu once talked about a special kind of bicycle that moves backward. If the steering wheel is turned left, the wheel would turn over and the bike moves backward instead of forward. This is actually related to the idea of artificial intelligence. There are several ways to learn this. One is experiential learning. For example, we may fall from the bicycle when we start learning to ride it, and we'll never forget once we've learned it. If we master the special bicycle, then we will not be able to ride the standard ones. We shall not bind the future with the thinking of the traditional driving.

Improving the Internal Strength and External Environment of the Unmanned Vehicle

After reviewing the history of unmanned-vehicle technology, let us now go deep into the "eyes and ears" of unmanned vehicles from a technical level. In terms of range of visibility, the unmanned vehicle has an absolute advantage over the naked eye. The human driver's safety range of visibility is about 50 meters (164 feet), while the unmanned vehicle can achieve over-the-horizon scanning observations of more than 200 meters (656 feet) through a variety of sensors, such as medium- and long-range radar and cameras.

However, the "eyes" of unmanned vehicles must overcome many problems that can be easily solved by humans. In order to allow it to judge the cars and surrounding road conditions from different angles, we need to teach the computer to distinguish cars, sky, and trees. Deep learning can play a huge role in this process.

In 2018, Baidu unmanned vehicle's accuracy rate was about 90.13 percent in the use of cameras to judge objects. In 2016, it was about 89.6 percent. It seems that there is only an increase of 0.53 percent, but, for the safety of unmanned vehicle passengers, every improvement is worth it. In judging pedestrians, the accuracy rate of Baidu unmanned vehicles reached 95 percent, and the accuracy rate of judging traffic lights is nearly 99.9 percent, with a future goal of 100 percent.

Of course, for unmanned vehicles, it is not enough to "see." To ensure safety, they must respond quickly, such as to brake whenever warranted. It takes human drivers 0.6 seconds from encountering an emergency to brake. But, it takes another 0.6 seconds for the brakes of the car to work, as the hydraulic system needs time to operate. In other words, average human drivers need 1.2 seconds to brake the car. Baidu's unmanned vehicle only takes 0.2 seconds from "discovering" an emergency to braking. If electric brakes replace hydraulic brakes, Baidu's unmanned vehicles will be able to brake urgently within 0.2 seconds, which will be 1 second faster than the average human. In high-speed driving, 1 second may save more lives.

In most driving circumstances, the radar, sensors, cameras, etc., of the vehicle must collect data in real time and feed it back to the Baidu Auto Brain server on the road, supplemented by GPS high-precision maps, to guide cars to travel in the best route.

Baidu Maps achieved GPS positioning at an accuracy level of 30 meters in 2015, an indoor high-precision positioning accuracy (for finding people or objects in buildings) from 1 to 3 meters, and a positioning speed of 0.2 seconds. In 2016, the accuracy of the high-precision map used by Baidu's unmanned vehicles was 10 cm. The error length in judging the road condition during actual driving was as narrow as the width of one painted lane line. Compared with GPS positioning, the accuracy was improved by two orders of magnitude.

It all comes down to the countless users of Baidu products, including those who use Baidu Maps APP. When the user accesses the service, it helps the unmanned vehicle to refresh the data and adds a little more "wisdom."

The key to the smooth development of unmanned vehicles is a reasonable technical layout. With the upgrading and innovation of technology, the problem of popularization and clinging to the past will be gradually resolved. This applies to the debate in the industry about whether an unmanned vehicle should use laser radar.

Due to the high cost of laser radar, some outsiders are pessimistic about unmanned vehicles, just as years ago many people did not rate "big-brother mobile telephone" highly due to its high price. Baidu is not only confident in price cuts but also forward-looking toward strategic investments.

At the end of 2015, the price of LiDAR (light detection and ranging) laser radar, commonly used for unmanned vehicles, was estimated 700,000 yuan (about $99,043). After half a year, the wholesale price of the same model has dropped to 500,000 yuan (about $70,745), a drop of almost 30 percent. With the maturity in the production process and the scale effect brought about by the development of the unmanned vehicle industry, the laser radar will have a greater price reduction. After all, the "big-brother mobile telephone" was as high as 20,000 yuan ($2,829, roughly) when it first came out, but now the domestic smartphones cost only about 600 yuan (about $85).

On August 17, 2016, Baidu and Ford jointly invested $150 million in the laser radar company Velodyne Lidar. Velodyne predicts that with one million orders in 2017 the unit price of the sixty-four-line laser radar used by Baidu's unmanned vehicles will drop to $500, $300 in 2020, and $200 in 2025.

The sixty-four-line laser radar is currently in short supply, and this investment ensures Baidu's sensor supply. But such a huge investment is not only for the temporary supply of devices but also to help make Baidu an industry leader, which will facilitate the development of the entire unmanned car industry.

To improve hardware-computing capabilities, Baidu independently developed a small cluster of forty-eight servers, and its computing power surpassed the Chinese supercomputer Sunway TaihuLight.

More than ten years after the start of Baidu Map, to the fruitful achievements of Baidu Brain's perception and decision-making ability and to the investment on laser-radar supplier Velodyne, it is not that Baidu chose unmanned vehicles, but that Baidu's professional logic has pushed itself to shouldering responsibility.

Feasibility on the technical level is still not enough. In order to accumulate road-test experience, Baidu's unmanned vehicles are struggling to run on the designated test sites at home and abroad. In California, Baidu is only the fifteenth company to receive an unmanned-vehicle test license. At the end of 2016, Baidu deployed more than one hundred unmanned vehicle researchers and engineers in California. In China, Baidu should also do its best to customize for China. After considering road facilities, pedestrian density, government (relevant departments) support, and even local weather, Baidu's driverless teams settled in Beijing, Shanghai, and Shenzhen for testing and initial commercialization. Cooperation agreements have been signed with Anhui Wuhu Municipal Government; Shanghai International Motor City; the Wuzhen, Zhejiang, tourist area; and Beijing Yizhuang Development Zone.

In 2016, in the closed test area of the approved National Intelligent Connected Vehicle (Shanghai) Pilot Zone, urban traffic scenes were simulated, with tunnels, boulevards, refueling and charging stations, underground parking lots, intersections, T-junctions, road circles (a.k.a. roundabouts), six smart traffic lights, and forty kinds of cameras. The testing area had centimeter-level positioning and WiFi coverage through the Baidu system and provided twenty-nine different scenarios for intelligent driving.

On March 22, 2018, the relevant departments of Beijing authority issued the first batch of temporary license plates for the automatic driving test in Beijing after a series of procedures, such as closed test-site training, automatic driving ability evaluation, and expert review. This not only means that Chinese autonomous vehicle companies can bid farewell to the transoceanic road tests but also represents a milestone in the process of automatic-driving legalization.

During 2017, Baidu launched Road Hackers, the advanced auto-driving artificial-intelligence model and will open Baidu auto-driving training data based on this model. The first phase contains 10,000 km of L3 data from highways, jug handles, and expressways of closed roads in more than a dozen cities. This is not only road data but also the driving habits of Chinese, which are more emotional and variable than machine driving.

The foregoing tests were just a sampling of Baidu's open automatic driving technology. On April 19, 2017, the Shanghai International Automobile Industry Exhibition was held. With a dazzling array of new cars on the market, the auto show was in full swing when a few technology and car reporters were attracted to an inconspicuous conference room in the exhibition area.

In this temporary, less spacious conference room, Qi Lu announced the Apollo program, in which Baidu would open its autonomous-driving technology and work with partners to create an autonomous-driving environment. The release of Apollo's open platform marked Baidu's system-level open process of artificial intelligence, which is the first system-level openness of autonomous-driving technology worldwide.

The insiders of the automobile industry immediately became keenly aware that the impact of Baidu Apollo on the auto-driving industry, and the entire automotive industry was tantamount to a targeted nuclear explosion.

The following story seems to confirm this judgment. From Apollo 1.0 to Apollo 1.5 and Apollo 2.0, Apollo's technology openness has been accelerating. More important, the process of product development, mass production, and internationalization of the entire industry brought about by the Apollo technology open platform is accelerating. At Present, Apollo has more than one hundred partners, making it the most open, complete, and powerful autonomous-driving environment in the world.

Different types of Apollo auto-driving vehicles are driving on the streets of Xiong'an, at the US Consumer Electronics Show, and on the stage of China's Spring Festival Gala. More important, the Apollo platform is using technology's openness as a link to create an unprecedented impact on the open community. This community, which is built with the concept of "opening up, sharing resources, accelerating innovation and continuing to win," has brought about the possibility of the Chinese auto industry surpassing others. We believe that autonomous driving should not be exclusive to few people but should be a standard feature of every car. Unmanned vehicles carry a lot of hopes of the Chinese auto industry. This is definitely not something that Baidu can do single-handedly. It requires the joint effort from the entire industry, including governments at all levels, automobile-manufacturing companies, scientific-research institutions, banks, and insurance companies.

Where Are Sophisticated Drivers Heading?

Although unmanned vehicles are far away from commercialization, they have already faced serious misgivings from all sides, which is a barrier apart from technology. Is the connected device on the unmanned vehicle invading personal privacy? Will serious consequences occur if hackers steal the data? How do we determine the responsibility in accidents involving unmanned vehicles?

People often ask an ethical question about unmanned vehicles: when a kid suddenly jumps out from one side of the road and the unmanned car turns the steering wheel, it may hit a pedestrian on the other side. How does it choose? The more advanced version of the question is "If there's one person on one side, and a group of people on the other side, how does it choose?" In this regard, the technical answer is usually to take the approach of braking and staying in the lane to slow down as quickly as possible and avoid tragedy. But the point is that even sophisticated drivers don't know the perfect answer for this ethical issue. For unmanned vehicles, humans should be more understanding.

Google's unmanned vehicle responded to various situations during the experimental period: a woman sitting in an electric wheelchair chasing a group of ducks on the road, a group of people lining up on the road for leapfrog, a man suddenly approaching and rolling over the hood of the car without warning, and even a man running in front of the car. Someone asked Baidu to comment about why Baidu's unmanned vehicle program has more than three thousand scenes and more than ten thousand scenarios. Baidu has no real statistics on this data, but these scenarios are obviously prepared to test scenes such as crashing into a person or a dog, at least to ensure that unmanned cars can make the best choices for the safety of traffic participants during the special situations that ordinary people can think of.

Imagine that in the future of IoT, when faced with an emergency difficult for human beings to decide, artificial intelligence is likely to give better solutions than humans due to its speed of response and the networking effect. The Internet of vehicles system will connect the unmanned vehicles into a whole, and each car has a big picture in the "brain" while driving. An emergency brake of a car will urgently notify other unmanned vehicles within a certain range to take corresponding measures to avoid rear-end collisions.

Although the public may fear that the unmanned car is too aggressive, engineers are worried that the unmanned car is too weak in getting along with people. In Google's experiment, when the unmanned vehicle was driving in the right lane, it found a sandbag in front, and then it tried to slow down and drive in the left lane. It thought that the rear bus would slow down and let it go first. Unexpectedly, the bus speeded up for passing while the car was changing its lane, resulting in a collision. In this accident, the unmanned vehicle obeying the traffic rules was "bullied" by humans. It is impossible to allow illegal behavior by unmanned vehicles. So how does the car get along with the flexible and clever human beings and how does it make decisions in the fuzzy areas of actual driving? Research on unmanned vehicles is concentrated on this problem. This is really much more complicated than a chessboard. It is still too early to worry that a defensive-driving unmanned vehicle will be actively harmful. These are hypothetical issues concerning the laws and regulations of unmanned vehicles, the regulation of auto-driving cars, and the determination of responsibility in accidents.

Currently, driverless technology has encountered many legal barriers in the European Union and some states in the United States. EU traffic laws stipulate that a car must be driven by a driver holding the appropriate driver's license. No one can take over the driving right for any reason. Only four states in the United States support autonomous vehicles on the road.

But, the difficulty at the legal level will not last long. In 2015, according to Google, "the National Highway Traffic Safety Administration understands the 'driver' of such a vehicle as an automatic driving system, not a person in the car." That is to say, it accepted Google's statement that Google's driverless cars will not have drivers in the traditional sense. In early 2016, the United Nations Economic Commission for Europe revised the Vienna

Convention on Road Traffic from the original "the responsibility of driving a vehicle must be undertaken by a human driver" to add "in full compliance with the United Nations vehicle management regulations, or the driver may choose technology; the transfer of driving responsibilities to autonomous driving techniques can be allowed to be applied to transport."

Because cars were invented before roads, traffic regulations came later. From technology to related legal systems, we must follow such a path. The development of autonomous driving is no exception. In recent years, as a member of the National Committee of the Chinese People's Political Consultative Conference, I have continuously proposed the auto-driving policy and the law-related proposals on the platform of the national Two Sessions meeting, hoping to promote the legalization of autonomous driving and win more space in the future development of the automobile industry. Although the legalization of China's autonomous driving still has not happened, the Chinese tradition of "crossing the river by feeling the stones" (or making reforms slowly) has made the unmanned vehicle companies feel at ease. The industry is not too worried about the obstacles of the law. After all, the law is always catching up with reality. In the era when the algorithm may replace the law, technology is the core driving force for development. Once the Chinese unmanned car starts, its acceleration will be very promising.

After all, unmanned vehicles are for people. The sophisticated driver of the unmanned vehicle continues to learn, and skills are improving day by day. Scientists dream that one day unmanned vehicles can eliminate road anger, congestion, and even smog because of fewer cars and higher car efficiency. The unmanned vehicle will plan the perfect path for each trip and accurately calculate the time on the road. Inside the car, we can rest, work, study, entertain, relax, and even travel. Making full use of the time on the road can reduce the difference between living in the city center and the suburbs, thus affecting the urban layout and land price, further influencing the work and life of young people. Humans will have more time and space to develop their own skills, improve their health, and make up for the lack of family companionship. Let's begin to image the various possibilities brought about by unmanned cars!

Of course, the flip side is that we may suffer from loneliness. More single-seat and two-seat unmanned taxis generated for efficiency and silent journeys may make us yearn for the warmth and camaraderie of carpool. Chatting with the driver could become a fee-paying service. But there will be fewer car accidents. Once a traffic accident occurs, it might make headlines. Compared with technological progress, the progress of human nature is slow. Car civilization is the embodiment of modern industrial civilization. On this land, from animals racing for survival in ancient times, to the hundreds of millions of cars today, and then to the unmanned cars in the future, it is the evolutionary path of life. Cars are not just cars; highways are not just highways. Civilization is "on the road," and it is endless. Know more, do more, be more—let the unmanned car make us better humans.

7

AI BRINGS ABOUT THE DAWN OF
INCLUSIVE FINANCE

It seems that financial markets are always characterized by great uncertainty. Therefore, finance is often considered as a paradise for adventurers. Financial predators smile at each other, set off the bloody hurricane of the market, and then earn excess profits from people's fear. But in fact, they are not always able to control the situation and instead often shoot themselves in the foot, and even suddenly quit the market.

The other side of finance is often like the silent flow of water. It can help people to improve their material and spiritual life or self-advancement.

Whether ups and downs or silent improvement, behind it is the handling and response to the complex logic of both capital and information flow. The sense of financial acquisition has increasingly become an important social proposition. How could financial services no longer be a "big players' game" and better for hundreds of millions of ordinary people?

Artificial intelligence may be the best tool to deal with such massive amounts of information and uncertainties. Its innovative breakthroughs and practices in the financial sector are bringing about the dawn of inclusive finance.

"New Interns" Joining the Agency

In 1994, the United States was in its golden age. It was the only superpower in the world, thriving and prosperous. That summer, the FIFA World Cup just ended in Los Angeles, and the flames of new technology were rising in

Silicon Valley. Netscape browsers soon became popular around the world. China just got access to the Internet through a full-functional 64K line. In an effort to contain the financial bubble, the Federal Reserve, under the leadership of Alan Greenspan, began to raise interest rates sharply, but bond markets did not realize that the Fed was entering a cycle of rate hikes. The global financial crisis did not come until thirteen years later; in 1994, Wall Street was flourishing, with countless people from all over the world pouring in to find their dreams.

Before the start of summer vacation, I received a letter of internship from a Dow Jones subsidiary; the job was involved with financial information-processing systems.

I studied information management during my college life. After graduation, I studied in the United States and obtained a master's degree in computer science at the University at Buffalo, the State University of New York. The internship was a perfect combination of information management with computers. For those three years, I worked with financial news every day. Then I participated in the design of real-time financial-information system of the *Wall Street Journal* online version and became a senior engineer in Infoseek, the internationally renowned Internet company. While dealing with financial information, I watched the commercial battle in Silicon Valley through the *Wall Street Journal* and started thinking about how to deal with the problem of information cheating. Soon after, I put forward the idea of "super-chain analysis" technology and applied for a patent, which laid the foundation for the future engine development.

At that time, I did not realize that one day, with the development of artificial-intelligence technology, the change of the machine's role in the financial system would be so fundamental today. It would penetrate deeply into various aspects, such as financial-information processing, data analysis, risk control, credit reporting, smart investment (robo-advisors), smart customer acquisition, and quantitative investment.

David Shawn, the king of quantitative investment, said, "Finance is a wonderful business of information processing." That's why Zhu Guang, CEO of Du Xiaoman Financial, said, "The most revolutionary changes will occur at least in the financial field, because artificial intelligence is the ultimate in the cycle of data collection, analysis, and processing."

The machine's penetration in various financial services is essentially the result of continuous improvement of financial-information processing capabilities offered by machine learning. They include integrated user portraits and construction of risk-control models in the field of credit and anti-fraud, mined investment-decision factors, and the matching clients and individualized investment portfolios.

With the continuous advancement of natural-language recognition and information database technology, even pure financial-information processing has undergone substantial changes. The intern walking toward us is a robot, and its distinctive way of working is first visually reflected in the generation of a financial-analysis report.

A financial-information system is probably the most complicated and boring information system. A share transfer agreement contains more than two hundred pages, and there are a large number of annual reports, semi-annual reports, research reports, announcements, feedback, and due-diligence results. We don't know how many industry analysts are working to finish reading the information before they make decisions. Maybe, it is not that they are not diligent enough, but that reading this information is unrealistic.

Yang Xiaojing, former general manager of Baidu Data Finance, said about the industry that during the 1990s, it took about ten hours for a fund manager to read the daily market research report, news, transaction data, etc., which is equal to the workload of two days. In 2010, after the outbreak of mobile data, it took about ten months for him to absorb the information generated on the market every day. In 2016, it took about twenty years for the same fund manager to read all the information on the market that day, which is equivalent to his or her entire career. Therefore, fund managers urgently needed the assistance of advanced intelligent technologies, such as Baidu's natural-language processing technology.

Today, much industry news and structured key information are processed by using Baidu Financial's intelligent financial-information analysis system, which structures all the key information or reads the annual report of listed companies and then forms an analysis report—within minutes.

In this entire process, the machine is equivalent to a junior analyst of a financial institution, or even an intern who undertakes all the basic work. The working logic of this machine intern is similar to the process of extracting keywords and recombining them.

The machine can instantly read massive heterogeneous data such as announcements, financial statements, official releases, social platforms, securities quotes, real-time news, industry analysis reports, etc., of all listed companies. Technical tools such as OCR (optical character recognition) are used on pictures and forms in the text. Then, the extraction of key organizational information is made, which basically finds relationships, such as the industry upstream and downstream relationship, supply-chain relationship, stock-rights change history, fixed increase and significant asset recombination, data cross-validation between multiple financial statements, etc., forming and presenting a knowledge map about these complex relationships.

One step further, once analysts select a template that meets the requirements and determine the subject, the machine can generate a basic report text. Before the final output, analysts can manually check the accuracy of the report and add their unique personal analysis and conclusions, so that a format standard—and even illustrated—financial analysis report is done.

This potential intern is obviously not going to stop at the stage of simply processing information. Since it has already passed through the hall and into the inner chamber, it will certainly go further.

Artificial Intelligence Makes a Fairer Starting Point

The robot at first opens the door to enter the core area of traditional finance—credit rating.

On the afternoon of July 18, 2016, Baidu announced an investment in ZestFinance, a US financial technology company, which also got an investment from Jingdong Group. Being preferred by China's two major Internet giants makes ZestFinance, a data credit company that only served one hundred thousand Americans, known more to Chinese people.

The Los Angeles–based financial technology company uses machine-learning techniques to assess the credit risk index of personal loans. Its founders are Douglas Merrill, a former chief information officer and vice president of engineering at Google, and Sean Budd, a former head of First Capital International Group.

In the United States, ZestFinance challenges the industry giant FICO, which accounts for about 99 percent of the US credit-scoring market and the credit-scoring market in most developed countries.

ZestFinance believes that "all data are credit data." Whereas FICO's credit score only contains dozens of variables of the lender, ZestFinance's model is based on massive social-network data and unstructured data. It contains nearly ten thousand variables and forms an independent credit score based on big-data mining. The efficiency can be increased up to 90 percent as compared to the conventional credit-rating system. ZestFinance claims to be able to analyze more than ten thousand pieces of raw information data for each credit applicant within five seconds and to derive more than seventy thousand indicators that can measure the behavior.

Prior to investing in ZestFinance, Baidu announced, at the Baidu Alliance Summit, that "artificial intelligence will have a transformative impact on finance, and it can truly upgrade credit information." Baidu emphasized that "now Baidu's finance unit can decide to make an educational loan within one second." Underlying the "instant check" is the big-data risk control based on machine learning, which is a small test for improving the efficiency of credit services and increasing the coverage of financial services.

In general, like traditional financial institutions, big-data risk control also results in two lists: whitelist-based for credit and blacklist-based for anti-fraud. The latter is often cloaked in mystery because of the purpose of anti-fraud. For example, Palantir, the world's fourth-largest unicorn, an artificial intelligence company founded by Peter Thiel, not only helps the US security ministry to fight against terrorism but also is recognized all over the world for discovering Bernard Madoff's Ponzi scheme after combining and fully mining records and data of the past forty years.

But, we are more willing to tell the story of artificial intelligence in mainstream financial risk control.

According to the data of the People's Bank of China, as of September 2015, 370 million out of 870 million ordinary people included in the credit system of the People's Bank of China had credit records; personal credit reports and personal credit score can be designed for 275 million people. This means that there are still about 500 million in China who do not have any credit history and who are blocked from the threshold of traditional financial services.

Relying on the huge data foundation and the image-processing technology realized by artificial intelligence, Baidu Finance and other enterprises are quietly changing the problem of online processing of risk control and dropping the previously superior financial services to tend to those in need with no credit history.

For example, Li Liang, who has previously studied indoor design for four years in college, recently started searching on the Internet for UI (user interface) courses, training schools, and paying tuition fees by installments. He hopes to enter a large Internet company after learning these courses. However, high tuition fees in educational institutions have become his first barrier.

Li Liang didn't know that there are still many people who are searching for the same keywords on Baidu at the same time, although at this moment they have not directly interacted with Baidu Finance. The group requirements of these people were collected in Baidu big-data risk control in the form of data and classified in a certain group portrait through machine learning, so they have the corresponding credit judgment.

After comparing several educational institutions, Li Liang finally chose a training school to learn UI and decided to try Baidu Umoney, which was recommended by the teacher, to pay the tuition on installment loans. He then completed the loan application by filling in identifiable personal information and taking a personal image in just a few minutes via a mobile phone.

Baidu's risk-control strategy system responds quickly. With the support of the user portrait and image-recognition technology, the information of Li Liang is collected, processed, and analyzed, and the data-field result is sent to the risk-control platform, so as to complete the credit process. After a short wait, Li Liang received the text message for the first loan approval in his life.

After a few months of UI course study, Li Liang decided to learn VI (visual design) to get better prepared for a future job. This time, he was pleasantly surprised; because of the good repayment record and stable consumption record, the machine has expanded and improved his loan quota and credit-payment scenarios.

More important, Li Liang's first credit record designed by the machine can help him to enjoy more comprehensive and better financial services in more financial institutions apart from Baidu financial system and bid farewell to the previous lack of a credit score.

Zhu Guang once said, "In our society, who loans to the young people with no favoritism and supports them in the critical climbing stage of their lives? Now, the answer may be 'machines.'" When the machine completes digitizing its financial services for people, nothing can stop it from surging in the financial kingdom.

Smart Night Watchman of Personal Wallet

Warren Buffett, who has always been less interested in technology investment, probably wouldn't have thought that someone would name a smart investment-consulting software after him. The software, which pays tribute to the investment guru, is Warren, a cloud-based financial software designed by smart-cloud investment company Kensho Technologies (*Kensho* in Japanese describes Zen's clearness, meaning to see the essence in a phenomenon). It basically analyzes the impact of specific events (from natural disasters to election results) on the market by using big data and machine learning and presents results through easy-to-understand knowledge maps.

The software initially shocked Wall Street at its release, and many even telephoned Kensho's founder, Daniel Nadler, and called him a "traitor." On Wall Street, no matter whether artificial intelligence or any other gorgeous technology is used, it is normal to make money silently, but opening it to public and universalism must be a considered treason and heresy.

The control and processing of financial information itself are viewed as a monopolistic business. Bloomberg and Reuters estimate that long-term monopoly of financial data has a market capitalization of $26 billion. More and more users of Warren are breaking this situation.

Another company, Hedgeable, was founded to overthrow the Wall Street monopoly. Its founders, Michael and Matthew Kane, are twin brothers who have served the world's top hedge fund, Bridgewater Associates, as well as Morgan Stanley, the most famous investment bank and another Wall Street giant. Because the twins were increasingly tired of Wall Street because it only serves the world's richest people, they quit to create Hedgeable in New York, trying to provide investment-advisory services to the general public with the support of artificial-intelligence technology.

If investment consultants for Americans are commonplace, then, for the Chinese people who have accumulated wealth for more than thirty years, the service providers need to be popularized in China.

At the end of 2016, Wu Xiaobo, a famous financial writer, conducted a "consumption survey on the new middle class" and found that the class, which includes about 180 million people, is generally anxious about wealth preservation.

The high-net-worth people have always enjoyed private consultants in financial institutions. But who will defend the wallets of "new middle-class" and ease their wealth anxiety?

Smart investment, which is also known as robot investment and smart financial management, usually refers to the process by which a computer, based on artificial intelligence and big data, provides users (of different risk preferences and investment requirements) with algorithm-based investment-management advice, to help investors to make personalized asset and investment decisions and to realize the optimization of personal asset allocation on different risk preferences and investment requirements.

Smart investment advisors generally follow certain steps. First, they understand the requirements of investors; that is, they clearly understand the key data of investors themselves and their family as a whole. For example, the investor's life stage, income level, historical investment experience, and preferences need to be considered. In general, the richer the investor and the finer the granularity of the portrait, the more accurate the understanding of the investor would be.

Financial management means wealth protection, long-term investment, and asset allocation; it is also a way of life planning. Therefore, machines must describe in detail the life pursuits of investors, such as buying a house, buying a car, studying, parenting, and pensioning, in order to match the corresponding investment cycle and examine the return-on-investment expectations.

Next, the machine will examine the investor's appetite for risk. Age, stage of the career, income structure, and expenditures all are used to determine the investor's threshold on taking risks. The risk preferences expressed in a face-to-face talk with investment advisors often deviate greatly from the investor's true thoughts. This requires investment advisors to have a professional and meticulous communication with investors, so as to gain clear insight into the real risk threshold, which is a costly process. However, machines identify customer risk preferences through big data and can dynamically adjust in real time according to market conditions, drawing on the investor's risk-preference curve, which greatly reduces communication cost, and the price investors have to pay.

After fully grasping the basic situation of investors, it is necessary to select the most suitable asset-allocation combination among various financial products according to specific customer characteristics. Therefore, while completing the investor's portrait, the machine advisors must also carefully understand the financial products, such as, the basic assets behind the optimal investment target—for example, asset characteristics, volatility, price to earnings, stability, and the correlation between multiple assets.

After understanding both sides, the machine must consider how to make a combination of matches, instead of a single match. This process requires strong computing power and efficient deep-learning algorithms. This is why a technology company like Baidu can cut into this field and swim like a fish in water.

Finally, there must be asset monitoring and control. As the market adjusts the asset portfolio, the machine needs to continuously update its investment plan to match user requirements.

The emergence of smart investment advisors has changed the way financial management agencies interact with customers. They can truly understand the investor's preferences, so that resources can be allocated through customized production. More important, with the help of artificial-intelligence technology, smart investment advisors can provide personalized and exclusive financial management solutions to the ordinary middle class at a low cost. Generally, there are no specific investment thresholds for smart investment companies. The management fee is only about 0.15–0.35 percent. The larger the user's investment amount, the lower the rate charged. This fee is equivalent to only about 10 percent of the cost of a human investment consultant. Some smart investment companies can reduce the cost to as low as 5 percent of the cost of human consultants, or can even do it for free.

In addition, machine advisors can effectively avoid human weaknesses. In the tactical asset allocation, such as stock investment, once general investors feel trapped, they can only wait for the price to go up. Once the income is obtained, they cannot get the money in pocket fast enough. Machine advisors cannot be affected by emotions. After setting profit points and stop-loss points, they can be executed automatically and strictly without any greed or fear.

Furthermore, machine advisors have unlimited energy. They can provide customers with personalized and modular 24/7 service through uninterrupted and reliable digital asset allocation and put forward service plans in time, based on the investor's demand preference and the market change. This ensures that machine investment advisors can communicate with countless customers on the phone or PC at the same time.

UBS Wealth Management, the world's largest and most famous private wealth-management institution, had about 4,250 wealth-management consultants around the world at the end of 2015, but it has to serve approximately 4.5 million individuals and businesses with a service coverage ratio of over 1:1,000. The efficiency of the machine will obviously be higher.

In recent years, smart investment has developed rapidly in the United States. In 2012, the US smart investment industry was not started yet. By the year 2014, the asset management scale of the smart investment industry reached $14 billion. Several companies, including Wealthfront, Betterment, and Hedgeable, have developed in the field of smart investment. Kensho, which provides investment-decision information services, is another smart investment company. KPMG estimates that by 2020 the asset management of US smart investment will reach about $2.2 trillion.

In China, smart investment companies that imitated foreign models began to appear in around 2015. During 2016, a large number of products with smart investment began to emerge. Although it seems that the market is hot, we have to admit that smart investing has been hyped. For example, some companies only recommend a certain fixed investment portfolio in the name of smart investment. In fact, they have long deviated from the original intention of smart investment in terms of risk allocation, investment objectives, investment ability, and willingness.

It requires comprehensive capabilities such as big data, machine learning, and financial insights to engage this industry, and the Chinese market environment calls for higher requirements. For example, the investment targets of foreign smart investment are mainly ETFs (transactional open index funds, often referred to as exchange-traded funds), which are passively managed funds. Correspondingly, the US market has nearly 1,600 ETFs, and the assets under management total $2.15 trillion. However, the total number of ETFs listed in China are about 130, and the total assets are nearly 472.9 billion yuan (about $66.8 billion, according to Wind Data Service, as of June 2016).

The shortage of index-type products has forced China's smart investment to introduce active-management funds. However, many changes in the active-management fund are unpredictable. For example, changes in fund manager or fund-company strategy result in a huge change in the fund's earnings, which is difficult to predict and simulate. This makes extremely high demands on the investment ability of smart investment companies.

In order to make a more accurate investor portrait, general smart investment companies will draw on the form of offline investment consultants and ask their customers to fill out the questionnaire. However, in China's investment and financial-management market, the investment group is dominated by retail investors, which have a greater speculative mentality than investment psychology. This easily leads to risk-preference distortions during the customer interview. For example, when investors find that some high-risk assets have good historical returns, they may ignore the potential risks and fill in a falsely high-risk acceptance in the questionnaire. On this issue, big-data companies, which can carry out user portrait including indicators such as "investment risk is above 100,000 RMB" and risk assessment rather than using general evaluation indicators, have a very obvious advantage.

The United States has a type of third-party organization that can bring together all the accounts of investors. As long as the user clicks on the authorization, the smart investment company can obtain information such as all the user's cash flow. For example, Pefin, a smart investment company, can manage all the accounts for investors under one platform, including savings accounts, bank consumer accounts, credit-card accounts, monthly payments, loans and investment accounts, etc. Pefin can use its analytical model to build a knowledge map of the investor's current financial situation in no time.

However, there is no such account consolidation agency in China. To generate an investor portrait as comprehensive as possible, we also need the machine-learning portrait ability based on big data, in addition to questionnaires filled by users themselves.

The biggest challenge is investor education. In a market dominated by retail investors, the main problem is how to convince investors that machines can carry out professional asset allocation and provide them with good financial management services. Moreover, in case traditional investment education has not yet been completed, the education process is difficult. Yuan Yue of Du Xiaoman Financial's Robo-Advisor team described the situation as skipping grades: "It's like you want to go to junior high school even when you haven't yet graduated from primary school."

The "black box" of machine learning has also increased the difficulty of education. Traditional investment advisers can explain the decision logic to the customer after making an investment transaction, but it is rather difficult for machine advisors to explain their "ideas."

In order to overcome this apparent defect, many smart investment companies choose to emphasize the professional factors behind their investment logic. An example is China Merchants Bank's MachineGene Investment, which highlights its comprehensive grasp of many fund companies and their fund managers as an important investment model factor. At the other end, Hedgeable has created an investment-advisory section that allows registered investment advisors, financial advisors, certified financial planners, certified public accountants, lawyers, insurance agents, and other key players in the financial value chain use the Hedgeable platform to serve their clients.

Of course, the development of the Chinese smart investment industry depends on the gradual clarification of future regulatory policies. If the smart investment industry is unable to provide direct investment services to individual investors' securities accounts due to license restrictions, then the best method is to export technical capabilities to institutions, which may produce a wider gap between institutional investors and public investors.

It is still impossible for a smart investment service to accurately say it benefits the public as a whole. It can only serve the middle class of a limited size in addition to the high net-worth group or serve as a technical exporter to an investment institution. This is determined by the zero-sum rules of the capital market. That is, when one person makes money, the other person must lose money. Until the machine discovers a better strategy through algorithms and procedures, the rational approach is to follow Wall Street's rule for "making money silently." If this strategy is widely used by the public, the rate of return would inevitably be greatly reduced and eventually fail. This is the paradox of smart investment.

In the case of Wealthfront, the chief investment officer is Burton Malkiel, author of *A Random Walk Down Wall Street*; the passive investment philosophy he advocated in this book is that since the market cannot be defeated consistently, then the individual should simply invest in it.

Wealthfront follows this philosophy and chooses an ETF, a passive investment vehicle that tracks the index, as an investment to achieve long-term and stable returns. But, as an open strategy, this also means there is no high returns. As a matter of fact, the wealth managed by these smart companies such as Wealthfront and Betterment is about $3 billion to $5 billion, which is far from the traditional, mega-asset management companies, such as BlackRock, that can easily manage trillions of dollars.

Kensho finally merged with Wall Street investment bank giants Goldman Sachs ($15 million in financing), GV (formerly Google Ventures), and New Enterprise Associates ($10 million in financing), etc. Its valuable data-processing technology was eventually closed to the public, being confined to the mysterious circle of Wall Street, and the revolution it spurred ended abruptly.

Some of the world's top asset-management companies such as Charles Schwab, Fidelity Investments, and Vanguard and international investment banks such as Goldman Sachs, JP Morgan, and UBS Wealth Management are also joining the smart investment field by investing in mergers and acquisitions or self-built platforms.

The big players at the investment table are waking up before the arrival of the intelligent revolution in the financial sector. In the future, maybe only a technology giant with strong artificial-intelligence technology and massive data can compete and cooperate with it.

Data Mining: The Key to Smart Investment

In November 2016, the US presidential election was in full swing. At the same time, another battle in the investment market was taking place.

Analysts of Credit Suisse found that CTA (commodity trading advisor, or a futures investment) funds with assets under management of about $330 billion, which specialize in both quantization and long and short strategies, were gradually turning to short positions. The short positions for US stocks soared to the recent highest level. On the other hand, long/short equity funds with assets under management of more than $215 billion set a new high for nine-month positions against US stocks.

Most of the investment transactions on both sides were carried out through algorithms and models by using the computers. The media defined it as a "robot war."

Since most of the transactions for quantitative investments are done by using computers through various models and algorithms, many people understand them as artificial-intelligence interventions in the investment market. But in fact, quantitative investment is just used to find the risk-free arbitrage opportunities in the market with the support of powerful computing. It can only be categorized as trading strategy, which mostly has nothing to do with artificial intelligence.

True smart investment is still data-driven. Regardless of the iteration of algorithm, creation, and the ingenious logical relationship design, the financial algorithm model still needs to be facilitated by a large number of multidimensional data sets that meet the requirement of the model, such as economic, social, and industry-specific changes, to verify the model's feasibility and precision.

An open big-data environment is crucial for smart investment or financial-information analysis. Because the cost of acquiring digital data from the physical world is extremely high, most of the companies do not have their own big-data resources, and there can be no intelligent investment analysis and decision-making to speak of.

Baidu and Google's large, multidimensional, big-data resources of search data and map data are rich resources for the mining of financial-data features from the Internet, which is different from the other traditional financial-system data. Together with leading artificial-intelligence technology, it has endowed search engine companies with the best resources to break into the financial-investment field.

With resources, there will be "miners." The alchemic process for big data is generally like this: After determining the data source, the entire network data will be integrated at high speed with mature big-data technology; then the system optimizes data-operation efficiency and maintains the integrity of network data. Next, advanced machine learning, artificial intelligence, big-data analysis, and other similar technologies analyze and process the massive data and explore the personalized characteristics of financial assets.

For example, each day Baidu generates approximately 20 million searches directly related to stock names or stock codes. The search volume of a stock tends to be highly positively correlated with the stock price trend, with an average correlation of 0.7 or more. The search volume represents the degree to which investors are focused on a stock. Assisted by public opinion, the information can help buyers and sellers determine when to enter the market and when not to. Searching data can yield a gold mine of information for making investment decisions.

If searching is a manifestation of subjective intent, then with sufficient data, people's potential requests and interests can be extracted. The special user structure of A-share market (mostly retail investors) and operational characteristics (customary speculation) can make search data an excellent indicator. Therefore, when the search data assists market selection, then we can effectively observe market changes.

Search data can be further explored and refined. For example, Baidu Map has already marked more than three thousand industrial parks and more than four thousand commercial districts in China. Observation of these industrial parks and commercial districts can help us to effectively judge the change in population flow in a region, business center, scenic spot, or even city.

The intelligent system can even refine the knowledge map of an enterprise and automatically update it in real time by machine learning. Therefore, with the help of rich data, it becomes an easy task for the investor to monitor the changes in the operation of the invested enterprise in real time with a broader view and more timely perspective and obtain in-time reference for investment decisions.

Traditional financial logic, investment logic, asset-management capabilities, and highly correlated real-time data will undoubtedly greatly improve the accuracy and forward-thinking of investment judgments.

In the financial sector, every 1 percent increase in efficiency and reduction in risk means huge wealth gains.

Three Layers of Smart Finance

We need to respect the market—its complexity is far beyond human imagination. Geniuses have created relatively simple and perfect models to help us to understand the financial world abstractly, but in the process of breaking down the models, something is lost. These models are relatively static, and as time progresses, the models may become less precise. In the research of Preqin, a London-based consulting agency, the benefits of typical systemic funds were not as good as those operated by managers.

The machine-investment battle before the US presidential election in 2016 mentioned earlier may have been exaggerated by the media. As Buffett said, "Investing is not a game where the guy with the 160 IQ beats the guy with a 130 IQ."

Regardless of the algorithms and models, we must respect the financial laws and investment logic. Volatility is not caused by machine investment but by changes in market expectations. Humans make insights and decisions behind everything, and this will not change, at least in the foreseeable future.

At present, many artificial-intelligence investment companies claim that their system's stock trading is completely detached from human intervention. For example, Babak Hodjat, a chief scientist and CEO of Sentient, which Li Ka-shing bought into, announced that "our system allows the fund to automatically adjust the risk level." Ben Goertzel, the founder of Aidyia, a hedge-fund company that analyzes the US stock market using artificial intelligence, is more confident about his system. "If we all die, it would keep trading."

But it is quite difficult to imagine a completely machine-controlled investment market. If most of the funds use artificial intelligence entirely, we may get results that we don't want. On one hand, under precise calculation, the rhythm and target of machine investment would be increasingly similar; the market fluctuations would get smaller and smaller, becoming extremely boring. On the other hand, since the investment objects of machine selection are more and more rigorous, the seeds of market imbalance are buried at the same time.

There is a scene in Liu Cixin's short story "The Mirror" in which everything in the future of human society can be accurately calculated and predicted. The development of human society will become stagnant, and ultimately civilization will be destroyed.

Such imagination may be gloomy, but it gives us the understanding that investment combines the essential desires of human beings and also includes the cultural characteristics that promote the continuous development of human society. The investment process is far more significant than digital growth.

Regarding the relationship between artificial-intelligence investment and the humans' investment decisions, Zhang Xuyang, vice president of Du Xiaoman Financial, explained his views at a seminar:

> Investment is a combination of technology, art, and philosophy. Baidu's cutting-edge big data and artificial-intelligence technologies can solve some technical problems. However, sometimes investment is artistic; otherwise, there would be no classical investment master like Buffett. Analysts will have different interpretations and methods for one market signal. For example, through Baidu big-data analysis, we found

that barbers and teachers have begun to get interested in stocks. For this information, some analysts may think this is a sell signal, while other may think this is a buy signal. This is actually a perception of the improvement of investment experience. In the process of judgment making, investing is more like something perceivable but indescribable.

In some fields, where our current artificial-intelligence technologies can make breakthroughs, such as image recognition, speech recognition, natural semantic understanding, user portraits, algorithms, and assisted decision making, clear signals can be extracted, and the machine can make judgments after self-learning. But, the most critical part in investment decision-making is unclear. It is something in meaning but not in language. Our current technology is powerless to understand such things, so as a result there is no way to make insights and judgments for people.

After AlphaGo defeated Lee Sedol, many people thought that the machine can make investments on behalf of the human. But in fact, the game of investment is completely different from Go. Go has a closed and well-informed game environment, whereas investment involves human's irrationality and has an open environment. So, in this case, technology needs to be upgraded a lot in order to replace the human and make decisions. This process needs at least ten years.

In fact, the artificial intelligence algorithms used for investment are similar. These intelligence algorithms are based on logistic regression and causal analysis. Later, with deep neural networks, there was a so-called gradient decision tree; finally, there was a genetic algorithm. However, the advancement of these algorithms has not yet gone beyond the scope of relevance analysis. Only some short-term memory can be realized. After all, the degree of response of the human brain has not yet been reached. Machines can perform well on some repetitive and recyclable investment decisions in the investment market. However, manual intervention is always required for markets with certain deficiencies, or for markets that are flawed in the artistic level.

I think it is still hard for machines to replace the human at the artistic level, at least so far. Behind our smart investment, we also need a team that can maintain an algorithm. The logic of this algorithm must be constantly adjusted to adapt to the arrangements in different investment environments.

The third level of investment is philosophy and self-discipline. That is, why do we make investments? We must have a rule to stop loss and make profit. In this aspect, machines may perform better than humans. In fact, it is easy to stop loss or profit through the machine, because people are inevitably influenced by mood swings, greed, and confidence. For example, people would always think, *I'm different from others; unlike the last time, this time I can escape with someone picking me up. I can make this shot because I know that although the bubble is big, I am confident I won't be the last one.* But, history is often repeated. And through the machine's algorithm, you can set it up to admit defeat, or stop profit at some point; there may be a process of thesis–antithesis–synthesis.

However, having machines replace people at the artistic level is difficult. Our smart investment requires a team-maintenance algorithm. The logic of this algorithm must be constantly adjusted to adapt to different investment environments.

In short, humans are still the most crucial determinant in the financial investment market, especially when the integration of finance and artificial intelligence is accelerating.

Just as a large number of physicists and mathematicians poured into Wall Street and brought revolutionary changes, nowadays, the cross-border flow and integration of artificial intelligence and financial talent is emerging in technical companies like Baidu.

"Robin said welcome back from Silicon Valley. So here I am," Wu Jianmin, who previously worked at Microsoft Research Institute, said about why he came to Baidu. He is mainly engaged in research on smart customer acquisition and other related research in Baidu Finance. He and more artificial-intelligence experts solve financial problems by technology. Zhang Xuyang, who manages financial management, and Huang Shuang are responsible for consumer credit, and other professionals from traditional financial institutions are those who raise and define questions.

This cross-border combination of talent is an inspiring innovation. We will wait and see how the convergence of artificial intelligence and financial service will revolutionize the financial-service industry in terms of identity authentication, big-data risk control, and smart investment. For individuals, artificial-intelligence financial instruments are necessary. Five to ten years from now, when we go to banks, securities companies, insurance companies, and other institutions to get financial management, credit, and financial services, artificial intelligence will be working in the background. Zhu Guang called the future state of financial technology "AI inside." We hope that our own technology, data, and capabilities will support all financial institutions in China to reduce the impact of financial uncertainty and to give full play to finance to help realize a better life and the dream of inclusive finance.

8

EVERY ENTERPRISE NEEDS A CHIEF AI OFFICER

A new civilization combining data and algorithms with personal life, social governance, and economic structure is coming to the fore. The evolution of this civilization is not based on paper but on the efforts of thousands of entrepreneurs and laborers, as well as balanced governance. This process is full of hardships and uncertainties.

Chinese entrepreneurs have heard the voices from the Internet of things and artificial intelligence, but the direction is unclear. Do companies have enough data resources? Which aspect should they start with? More important, is there enough reliable artificial-intelligence talent?

For technical enterprises, this is not a difficult problem. The artificial-intelligence upgrades of leading-edge industries and high-end enterprises have already started. Many large enterprises have already begun the layout of intelligence-related products, such as smart home appliances. However, because technology is rooted in human life, artificial intelligence will only prosper when the majority of enterprises champion it. Those organizations should enjoy the opportunity to profit from artificial intelligence.

Traditional state-owned enterprises are undergoing various demanding reforms. Taiyuan Railway Bureau has the largest volume of freight and the most advanced technology in China. After the railroad and electrical-engineering upgrades for the Industry 1.0 and 2.0 eras, it now faces new upgrading challenges for the Industry 3.0 and even 4.0 eras. In the face of heavy traffic, business processes must be optimized to improve operational efficiency.

According to the analysis of artificial-intelligence scientists, the time required for a train to travel from one city to another city is relatively fixed. The transit efficiency is affected by other variables; when a train arrives, the goods are transferred from the train to the warehouse and then from the warehouse to another train. Reduction in the transit time would greatly increase logistics efficiency. Baidu cooperated with Taiyuan Railway Bureau to collect logistics data, place the data on Baidu cloud, and train with several models to accurately predict the arrival time of the train, the future storage demand, and the future capacity demand, reducing transit time by 50 percent.

Although it is not easy to have intelligent innovation in old state-owned enterprises, the bureau has accumulated technical teams and resources over the years. In contrast, we are more concerned about how small or medium enterprises (SMEs) can keep up with the wave of intelligence. We can say with certainty that artificial intelligence needs to be inclusive for SMEs. There is a large demand for artificial intelligence in both commercial and consumer markets. Ordinary enterprises, especially SMEs, have considerable opportunities in the intelligent age. Time waits for no man.

Who Will Break Through the Bottleneck of Product Upgrades?

The intelligent revolution must be initiated in leading disciplines and in leading companies and will also widen the gap between enterprises in the short term. However, AI is open, and it can be learned if companies work hard on it. In this process, we believe that traditional enterprises especially need guidance from professionals in the AI field.

The *China Manufacturing 2025* action plan, which was released in 2015, pointed out that "manufacturing is an important cornerstone supporting China's economic and social development and an important force for promoting global economic development." The report continues:

> Every country is ramping up in the scientific field and in technological innovation and promoting new breakthroughs in three-dimensional (3D) printing, mobile Internet, cloud computing, big data, bioengineering, new energy, new materials, etc. Smart manufacturing such as smart equipment and smart factories which are based on information physics systems is leading the transformation of manufacturing methods; network crowdsourcing, collaborative design, large-scale personalized

customization, precise supply chain management, full lifecycle management, and e-commerce are completely reshaping the industrial value chain system; smart terminal products such as wearable smart products, smart home appliances, and smart cars continue to expand into new areas of manufacturing. Industry transformation and upgrading, innovation, and development have ushered in major opportunities.

Intelligent upgrades are not exclusive to the industries involved in *China Manufacturing 2025*; it can be applied in almost all enterprises. Theaters and cinemas can optimize ticketing through intelligent learning of ticket-purchase rules; small supermarkets and small shops can analyze visitor routes through store electronic sensors and third-party data (such as Baidu Map passenger-flow data); traditional news outlets can innovate information production and push processes through the access of intelligent flow.

The people's understanding of upgrading is often limited to materials, processes, etc., but artificial intelligence brings a new dimension in this regard. In the case of household items, artificial intelligence has created many new scenes in this field: even if for just a curtain, someone has developed an intelligent solution that allows the curtain system to record and learn the habits of the owner in order to open and close automatically at the appropriate time to adjust the light in line with the owner's habits. Smart toilet seats are not heated all day, but only according to the household commuting schedule. Aged-care facilities can learn the living habits of the elderly, adjust the status of each home appliance, and even send an early warning signal to their family when an abnormal daily routine occurs. All we need is more creativity.

Historical Experience: The Brilliant Era of the Chief Electric Officer

As most know, during the dawn of the human industrial era, the promotion of textile machines and steam engines had suffered fierce opposition.

In the United Kingdom, the low-cost products produced by textile machinery overwhelmed traditional handmade textile industry, so industrial owners and workers opposed the machinery, setting off the so-called Luddite movement. The spinning jenny's inventor James Hargreaves moved because his neighbors damaged his machines. However, the spinning jenny eventually became widespread and helped the British rule the global spinning industry.

The steam train was not as fast as the carriage at first and was mocked by coachmen.

In the era of the electrical revolution, history was repeated again. For example, Guglielmo Marconi developed the earliest radio device in 1895 and successfully used his device to conduct long-distance Morse-code communication experiments. He founded a radiotelegraph and signal company to promote the commercial use of radio. However, due to a conflict of interest with a submarine cable company, his attempt to set up a wireless office in Newfoundland was rejected. However, at that time, the modern market system and technical preferences of the United States had been initially established, so the radio developed very quickly.

The electrical revolution is similar to today's intelligent revolution in the sense of basic materials. Unlike steam power, which cannot be transmitted over long distances and have a unified layout, electricity is an infinitely flowing universal energy source, just like the mobile Internet, a basic resource that allows users to access it at any time. "Electricity plus industry" was just like today's "Internet plus industry," which has overturned countless traditional industries.

Of course, current and intelligent flow can only be metaphorical. The former is the flow of electrons and the latter is the flow of bit coding—not the same thing. But this analogy can help us to understand the key problem. Let's compare the enterprise upgrades of the two eras.

When electricity flows to millions of companies, many companies seek upgrades. Although the companies are not as resistant as those during the steam era, there is still a sea of troubles. The power system more than a century ago was very complicated. People needed to make reasonable choices between DC and AC, different voltages, different levels of reliability, different power interfaces, and prices. Until today, countries have used different voltage and interface specifications (sockets). Dealing with different power companies is a technical job undertaken to save money. Similarly, companies need to select a professional Internet technology outsourcing company, or else they will fall into traps, because there is a dazzling array of choices, from programming languages to system architecture.

As far as the company's own business is concerned, it was difficult to find out how to use electricity to obtain the best benefits: should it be necessary to install lights for enterprises, or should electric motors replace the gas turbine? As a result, many companies created the position of vice-president of electricity to help organize reforms to ensure that for each function the company took electricity into consideration in its objectives, including installing electric wiring, purchasing power equipment, renovating existing equipment, and even transforming the company's business process (e.g., lights make night shifts normal).

Today, one might not know how many mind-blowing electricity-related products were available at the time. That's just like the steam-engine era; many steam devices were invented and disappeared later, which can only be found in steampunk cartoons today.

Chief electric officers needed to discover what values electrical products could bring to the company. The officer's vision should not have been limited to electricity as energy but also include electronic products. For example, people needed an alternating-current switch to control the output current based on the input voltage, which was created by extensive use of vacuum transistors in the 1950s, before the vacuum triode industry started. At the beginning, the low transistor production resulted in a poor quality control and a high price of $20 dollars for each piece, while a vacuum triode only cost $1. Which should companies have chosen? That depended on the vision about the future. Eventually, vacuum transistors were replaced by vacuum triodes because of their low energy consumption, long life, and small size.

As power systems developed, the role of chief electric officer disappeared. Every company that had a chief electric officer promoted the penetration of electricity into the overall production system and contributed toward the standardization of the national power system.

More than two decades ago, the same was the case for the Internet. The Internet had left the military laboratory and deeply penetrated enterprises, colleges, and even families. Networking is a complicated process, and it led to the job of network technical director.

In this era of Internet-for-all, many companies still choose network outsourcing services. From a resource-saving perspective, this is reasonable. Not all companies need to support expensive technical teams. It is convenient for companies to buy Internet services like buying electricity.

But the purchase of Internet services has two levels: One is infrastructure (IaaS)—access to the backbone network through service fees, which is a very standardized level of service. Another level is very individual—customized

Internet product services (PaaS and SaaS) on the basic platform, such as cloud servers, database systems, office collaboration tools, financial management tools, app programing, etc. This level is still more complicated for operating in traditional enterprises. When the company reaches a certain scale and still relies on outsourcing for technology, they may get lost in future development. So, the CTO (chief technology officer) is indispensable for both technology companies and traditional companies. The CTO in many small companies does not manage a large team. His or her job is to deal with outsiders on behalf of the company, to decide what services to choose, and to supervise the implementation process.

Therefore, excellent CTOs not only manage and operate the network but also dig into the value chain of all departments, products, and networks of the company and track the latest technological developments. They need to understand products and business, predict the long-term technical strategic direction, and constantly follow up on business and technology development; otherwise, they will be replaced by engineers or product managers, just like the electric power officers. (However, in most industrial enterprises, remnants of CPOs remain: electricians.)

Embracing the Smart Force

Artificial intelligence should become open like the Internet. The open artificial-intelligence service will become a kind of flow, as convenient as the electric current for users to access.

Li Zhifei, an expert on artificial intelligence and an entrepreneur, said during the Wuzhen Internet Conference in 2016, "We must clearly notice that in the short term, artificial intelligence cannot be compared with people. And, today we must make some practical applications, such as cloud interaction, natural-language understanding, computer vision, and other application scenarios—for example, in-vehicle devices, wearable devices, etc., to be used by people—not just discuss them in media or movies."

This coincides with Baidu's attitude. Foreign companies such as Google and Tesla have created a lot of interesting, smart public-relations projects. Both Go and QuickDraw have greatly increased the public's interest in artificial intelligence. Maybe only engineers on the West Coast of the United States have such a relaxed mindset to do something seemingly easy. Chinese artificial-intelligence companies with more responsibilities may not be so relaxed, and many peers have focused on developing enterprise-level applications, rather than publicity. According to reports, Google has split the once eye-catching business of the robot and unmanned vehicle from its main business and no longer unconditionally supports the two projects. What about Chinese companies? Can we promote industrial-grade development of artificial intelligence?

Baidu pursues the application of artificial-intelligence technology in enterprise upgrades, bringing improvements in the entire industry, rather than just creating some stories. We believe that artificial intelligence is to supply enterprises with a steady stream of "original forces." For example, the dataflow platform of Baidu Map and the voice-recognition services available for access are the most important.

Baidu XMind, a derivative product of Baidu Maps, can make a crowd portrait of resident population and show passenger flow, living-area price level, office-building density, and even the direction of crowd at a certain intersection. In cooperation with many chain stores, by the search of WiFi through the mobile phone, a positioning accuracy with a distance of less than 30 meters can be achieved without GPS. As long as the user's mobile phone receives the WiFi signal of the chain store, it can be viewed that the user enters the store. This is used to evaluate the market capacity and estimated turnover for the store. This function can also provide scientific advice for site selection of real estate, cinemas, etc.

Wu Haishan, former scientist at Baidu's big-data lab, explored the spatiotemporal data and applied artificial intelligence not only to commercial site selection but also to enterprise sales. For example, a catering company wants to issue one thousand coupons to one thousand users, but how does it ensure that these one thousand people had not visited the store before and already intended to buy? We can use machine-learning methods to create a model based on the characteristics of old customers. Which users match the consumption habits from the database according to the model of preferring to stay at home or going to clubs? What is the per-capita consumption of their favorite food and beverage outlets? Do they prefer traveling abroad or in China? Do they go to work by subway, car, or other means? By arranging the similarities between the old and new users, we choose the closest one thousand people and send the coupons to them; the success rate will obviously be much higher that way than with random flyers on the street.

Every enterprise needs a leader in artificial intelligence if it wants to find the right smart streaming service. In order to take advantage of artificial intelligence, traditional enterprises must understand what artificial intelligence can do and how it affects the company's strategy. How should the company organize the leadership team in preparation for the possibly disruptive introduction of AI?

From CTO to CAO: The Person Who Leads the Enterprise Upgrade

Every company wishes for someone to lead technology. Today, general companies have CTOs (chief technology officers) or CIOs (chief information officers). In response to the smart revolution, Andrew Ng wrote in November 2016 that every company needs a CAO (chief AI officer). So, where would a CAO come from? What is the relationship between CTO, CIO, and CAO?

With the full penetration of the Internet, the CTO came into being. Different from the former chief engineers, it is a role responsible for the Internet informatization wave. In many enterprises, junior technical officers often build an intranet platform and manage software and hardware. However, in the new era, not all CTOs have insight about intelligent business. For example, in some business schools, the CTO's responsibilities are limited to building networks, office systems, etc., based on demand, without gaining insight into and leveraging the data gains of the business itself. Many business schools sell course cases, which could be intelligentized to track and analyze customer subscriptions. Finance is closely related to data, but many financial institutions are not able to digitize themselves.

Many companies' Web 1.0 (static web pages) informatization and dataization are far from satisfactory, so some people have suggested that enterprises need a CDO (chief digital officer) or CIO. The CIO is an upgraded type of CTO to help companies organize information. The CIO thinks about how to improve the efficiency of the company's information transmission, how to share information rather than the repeated generation of similar information in various departments. According to some surveys, quite a few CIOs are nontechnical because they need to understand business and as well as technology.

The CDO must possess the ability to gain deep insight into the meaning of data, such as intuitively grasping data feedback through data-visualization techniques, discovering hidden value through data mining, and transforming inert data into active data, fueled by optimizing enterprise-data architecture.

In the past, the general IT department was usually focused on selecting, implementing, and integrating ERP systems; managing the company's servers and networks; and using computer equipment to train all new employees on internal software, processes, and internal business operations. Specific projects include employee information management, especially analysis of personnel recruitment and mobility; wage and cost management; insurance and benefits; accounting computerization; developmental project management; server management; storage; enterprise communications; marketing automation; customer-relationship management; customer-demand analysis; enterprise knowledge accumulation, etc.

Today, however, large ERP systems are being replaced with specialized SaaS products that are hosted by the cloud. These applications have a consumer-friendly user interface convenient for buying, using, and deploying. Companies will no longer need to spend several weeks of searching to determine the best payroll software and then months to implement it. For example, in addition to Jira (project and issue-tracking tool), project management can be developed by third-party office collaboration software. Server management includes the Amazon network service, Baidu Cloud, Alibaba Cloud, etc. Enterprise storage can be accomplished by Baidu Cloud Disk, and for enterprise communication, QQ for business, DingTalk, and others could work.

Information-technology services have been streamlined. Implementation of large SaaS tools still requires some integration, but trends over the past few years have shown that integration is less difficult and enterprise applications are being simplified.

In the past, CIOs were responsible for establishing a technical team. However, a skilled workforce can adapt to demand; companies can hire freelance computer programmers to deal with small tasks or look for mature "instant" expert teams through outsourcing. Various online project-management tools and resources facilitate outsourcing services. This aspect of the CIO's original responsibility is diminishing.

Chief electrical officers have long gone; will CIOs disappear too? Take Baidu as an example. Once we set up the job of mobile officer when mobile Internet was hot. The mobile officer was responsible for thinking and planning the mobilization of Baidu's service, but with the popularity of mobile Internet thinning, this position has been nullified.

CIOs familiar with their work will shift their business focus in a timely manner. Although third-party IT services are becoming more standardized and easy to access, the customer requirements are constantly changing, not as standardized as the company's own IT basic needs, and even less as standardized as electricity and energy. CIOs must respond to changes in information from customers and adjust corporate information services.

The future depends on the vision of CIOs. A new task is to organize complete enterprise data and develop smart strategies, no matter whether the position is called CDO or CAO.

With fewer tasks at hand, CTOs and CIOs have gained more time to devote to researching smart technologies. Third-party IT service is still universal, and companies need to pursue personalized customization. The same outsourcing company with different product managers of different capabilities will yield products of different quality. CIOs or CAOs must observe their own data and look for data resources that are not fully utilized yet.

The information that CAOs must deal with far exceeds financial information and includes production information, organizational information, and even employee-movement and position information. Think about why express companies need to know a delivery person's movement track. CAOs will consider collecting their movement data, using mathematical methods to analyze, suggest optimization, and improve delivery efficiency.

In the eyes of the CTO, data may be a technical by-product, but it is a strategic resource in the eyes of the CAO. He or she needs to spend a lot of energy to think about customer requirements and a complete user chain to find the ways for improving the company's products and services and try to interact with customers of all products.

For example, in the aviation field, China Southern Airlines Company, as the airline with the largest Chinese transport, the most developed route network, and the largest yearly passenger traffic, was the first airline company in China to join the Baidu Xingyun travel big-data platform, to transform "large" into "flexible." It utilizes the Baidu's location-service products, data, and market resources (such as Baidu Maps), to provide services to its passengers, including airport navigation, intelligent transportation from terminals to downtown, travel planning, dynamic guidance, etc. It leverages Baidu's technological advantages in cloud computing and big data for building a big-data-analysis model based on geolocation services; providing decision support, such as planning and designing, statistical analysis, and market monitoring; and creating travel services that are characterized by full process, high quality, multicontent, efficiency, and convenience.

From the Internet to big data, and then to artificial intelligence, these are not only conceptual changes but also cognitive and substantive changes, from form to body to soul. It is important for CTOs and CDOs to have an idea about the artificial-intelligence field.

CAOs will collaborate and innovate with different departments of the company (human resources, sales, marketing, products, etc.) to connect and integrate different departments. For in-depth data integration, interaction, and mining, more mature methods and tools will be needed. Currently, access to the artificial-intelligence flow is also coming up.

What Do Chief AI Officers Do?

To introduce mature machine-learning methods, CAO officers need to turn data into training materials, shape automation mechanisms, and then find and explore valuable management models and customer models, or sensibly introduce third-party artificial-intelligence flows to support business development and find a suitable innovation cycle: data→algorithms→ knowledge→user experience→new data, creating business value. This is what CAOs do.

CAOs faces many uncertainties. What criteria should they follow when looking for artificial-intelligence services? The CIO knows how to choose the right third-party, as individual users care more about computer-graphics or USB (Universal Serial Bus) interface standards. People often ask Baidu, "Is there any industry standard for artificial intelligence services?"

Andrew Ng believes that artificial intelligence is not very mature at the moment, and there is no unified service standard. Every company is working on it. At this level, the CAOs of each company are very important, because they also take responsibility for exploration. However, stable, convenient, and easy-to-use engineered products must be developed in the right direction and contribute to the standardization of artificial intelligence. For example, developers can easily use interfaces on the artificial-intelligence platform, enjoy stable intelligent stream output, and support enterprise operations and entrepreneurship, just like people create apps on mobile phones and development platforms. Baidu Brain is such a platform.

This round of artificial intelligence has caused a craze. One reason is the lowered threshold. The principles of deep learning, especially those related to algorithms, are similar. It is strong accumulation, rich data, and sufficient training that count. Just like countless programmers use Java for programming while their programming skills differ, so we can only try to choose the best.

For deep-learning neural networks, adjustment of neural-network parameters is extremely complicated and variable and requires optimization according to different business scenarios. Although the basic ideas of the algorithm are similar, the development methodology is different, so there is no unified standard. Therefore, training for neural networks is also called "alchemy." Andrew Ng's new book *Machine Learning Yearning* is a manual for "alchemy"—not about how good it is, but about the problems and solutions encountered in the process. Maybe deep experience and pragmatic approaches are the standard.

It is irrational to expect every executive to fully understand artificial intelligence. But if the industry can generate mass data, there is a great opportunity to use artificial intelligence to convert the data into value. For most companies with data but lacking deep artificial-intelligence knowledge, hiring a CAO or VP of AI is already an emergency, and some CDOs and forward-thinking CIOs are actually playing this role.

Companies without CAOs Will Be Obsolete

Suhail Doshi, CEO of the famous data-analysis company Mixpanel, believes that "machine learning is not to prove something; it exists to make some high-quality predictions on specific data, behavior, or patterns. The job of the algorithm is to enable us to achieve the goals more effectively and precisely, rather than telling us why."

However, we believe it is important to ask why. We have found in many enterprise applications that some entrepreneurs hope that deep-learning networks and data mining can immediately bring a substantial increase in profits, while ignoring the laws of machine learning and the learning patterns of enterprises themselves.

When the enterprise is large enough, and the business and data are complicated enough, then its operational logic is often vague, which is unlike the description in the company manual, and even enterprise managers find it unfamiliar. Machine-learning technique has the ability use backward logic based on data. This backward derivation capability can provide a way for business operators to observe the nondominant logic of the enterprise. For example, the analysis of the relationship between email transactions, order time of e-commerce users, and time period for news websites to publish information all can help companies understand which aspects they should optimize.

Enterprises are eager to step into the era of artificial intelligence from the Internet era. Currently, CAOs need to prioritize to optimize financial information or production capacity, or self-advancement or sales expansion.

In external business, CAOs can bring significant value. For example, consider an automated customer-demand analysis system that can be established to immediately submit order information and production information to customers for reference. But internal management is also very important. For manufacturing enterprises, for example, the informatization and dataization of production management, material management, quality management, and design change need to be integrated. After dataization reaches a certain point, then artificial intelligence will become a reality supplemented by algorithms and development.

In the computerization→informatization→intelligentization cycle, if the internal mechanism is not smooth, then the intelligence will lack foundation and become unsustainable. The internal cultivation and external expansion of enterprises will be highly unified in the artificial-intelligence era.

Different from the external input of current, in enterprises, intelligent materials exist in the form of "stream," which is generated in every work link of employees and machines. Every employee working on a computer generates a stream of intelligent data. Most companies fail to recognize the value and let the data evaporate.

The internal strength of an enterprise depends on each employee; factors such as personal information sharing, knowledge management, recording and optimization of work, and operation habits count.

For example, editors of online news media are promoting articles every day. Some online media have thousands of editors, each doing his or her own work. The intelligent system will detect and record entry operations and find out which parts require too many mouse clicks and which may indicate irrational design of the entry system; optimization suggestions for the entry system are proposed accordingly.

Andrew Ng believes that traditional enterprises especially need to rely on CAOs to better understand cutting-edge applications and to upgrade their own business. Of course, in the face of many possible choices, the CAO must first seize one department or one business to make a breakthrough and form a model to attract another department's interest. After all, enterprise intelligentization is a business of innovation and extensive participation.

CEOs must give authority to CAOs to play their role and even personally advocate the company's first intelligent business. Without the support of CEOs, the artificial intelligentization of enterprises will be very difficult. Some initiatives may not be understood until later. It would be a great misfortune if we miss them.

The Quality of CAOs

Is the CAO, as someone who deals with the most challenging deep-learning techniques, particularly rational or sensible?

Unlike general engineering technology, artificial intelligence mimics human thinking. Scientists emphasize that human-machine integration is the future direction of development. Similarly, smart scientists and CAOs are not one-dimensional adherents of mechanistic philosophy, but they have a comprehensive interest in all kinds of knowledge. Andrew Ng himself is an example. He likes computer science, anthropology, sociology, and pedagogy. To let more people receive the education of prestigious schools and maximize the reuse rate of classroom education, he cofounded the Coursera online educational program and remains a cochair of the board of directors to date. In addition, he is also very interested in psychology.

The CAO may be an interesting science geek such as portrayed in *The Big Bang Theory*—good at discovering opportunities that others ignore. He may also be like Sherlock Holmes in the data field—adept in seeing the truth. But first, he is a person with "data sensibility." He knows how to use data; more important, he can find data where others can't.

In the course of his work, Ya-Qin Zhang found that many companies had bought lot of servers to collect big data in the past few years, but they did not know what to do with data. Even for artificial-intelligence companies like Baidu, a lot of data can't be instantly used, but it may become useless in the future if it is not used in the moment. So, CAOs must play a role in deciding what to do with a huge data collection.

Walmart's shopping-basket analysis is a typical data-mining application, and data analysts want to find out the correlation between the purchase objects on the customer shopping receipts. After analysis, they found that in addition to the obvious relationship that milk and bread are often bought together, there are many new correlations that had not been discovered before. For example, baby diapers are often purchased with beer; later, they learned that young dads tend to treat themselves when buying baby products, so they bought beer at the same time.

Therefore, Walmart stores put beer close to the diapers, and the sales were significantly increased. Walmart's data is limited, with the only user data that of shopping receipts. If the data is huge and can automatically adjust the daily shelves, it becomes the prototype of artificial intelligence. Both Amazon's and Alibaba's e-commerce big data have reached a certain level, so they all provide a stage for artificial-intelligence scientists.

This example once again shows the mindset of CAOs, the relevant thinking we have repeatedly emphasized. The connection between milk and bread and between beer and diapers is clearly not a certainty or causal relationship, but a strong correlation expressed in probabilistic form. To explore value from the data is the quality of good CAOs.

The Second Layer: Beyond the Data, the CAO Needs to Have Scene Awareness and Understand Scene Calculation

In the future, the concept of business will become blurred; "scene" will become the core of operation, which means that the product is no longer categorized by business, such as news app, e-commerce app, etc. Now we can import all the services needed for one scene, such as an app for mobile payment; we can also consider various payment scenarios for it. If it allows for movie-ticket purchasing, we must guess what scenario the user is in, according to user's behavior data. In addition to watching movies, does the person need to read film reviews or buy popcorn? Enterprises need to replan according to the scenario.

From a perceptual level, the sensitivity to the scene reflects the CAO's appetite for life and humanistic feelings. Products for elderly care and the blind all come from this kind of feeling. In 2016, video-related enterprises developed rapidly, and someone programmed a smart editor that can automatically edit video. If the user enters keywords, such as "aircraft carrier," the software will roughly cut out the relevant clips in the video. This is an extraordinary application of visual-recognition technology, which is derived from the developer's sensitivity to the user scene.

Unlike general scientists who are accustomed to nonutilitarian exploration and discovery, the CAO needs to be more utilitarian; that is, he or she needs to have an economic sense to help companies discover value. This is reflected in optimization processes, accelerating response, saving time, and liberating people from repetitive mechanical labor. Artificial intelligence allows people to have more time for more meaningful things and to set aside time for human high-level mental activity.

Beyond such acumen are specific machine-learning operations and intelligent system-building capabilities. Based on the experience of some successful artificial-intelligence teams that he has led and supported, Andrew Ng listed some characteristics of successful CAOs:

- Good technical understanding of artificial intelligence and data infrastructure. For example, they have built important machine-learning systems. In the era of artificial intelligence, data infrastructure (how you organize your company's database and ensure that all relevant data is securely stored and accessible) is important, even though data infrastructure skills are relatively common.
- Cross-functional working. Artificial intelligence itself is not a product or business, but a fundamental technology to help the existing lines of business and create new products and lines of business. Therefore, it is important to have the ability to understand and work with different business units or functional teams.
- Strong entrepreneurial skills. A few years ago, it was unrealistic for artificial intelligence to create opportunities to build new products such as autonomous cars and communicable agents, which only exist in science fiction. However, entrepreneurs are often thinking about breakthroughs that allow them to create value updates from 0 to 1. A leader who masters entrepreneurial skills will increase the probability that a company will succeed in creating such innovation.

- The ability to attract and keep artificial-intelligence talent. Such talent is highly sought after. Good CAOs need to know how to keep talent, such as projects that talented workers focus on, opportunities to develop skills for team members, and so on. Use the project to educate others, instead of doing things in person, so as to form a training cycle mechanism for talents.

Andrew Ng believes that qualified CAOs should have experience in managing artificial-intelligence teams. Artificial intelligence has evolved so fast that they need to keep up with changes, but it is less important to ask them to be at the forefront. More important, they should be able to work with others across functions. Independent technology is meaningless—technology needs to serve people. The relationship between machines and people is very important. The best artificial intelligence requires an understanding on both technology and people. Artificial intelligence is a story between people and machines. CAOs must be good at machine intelligence as well as ways of the world.

CAOs must become evangelists and guides, giving the entire company an interest in artificial intelligence. They will be the corporate idols of the new era, just like the product master was the hero of the Internet age (think Steve Jobs). Such a hero will have knowledge, product and management thinking with regard to artificial intelligence, and appropriate human sensibility.

It is likely that in the near future companies that do not have CAOs or access to smart flows will be considered obsolete. From CTO to CIO, CDO, and CAO, it is an uphill process of cyclic rise. In a way, it also reflects the difficult process of self-renewal and self-evolution.

A Concern

Lack of talent is a common problem at present. Companies have ferreted people out from colleges, and intelligent scientists have left for enterprises to realize their dreams. Ng's good friend Yoshua Bengio, one of the top ten artificial-intelligence (deep-learning) scientists in the world, is worried that with many scientists working in enterprises, fewer of them will be engaged in in-depth education in colleges and universities, which will reduce the output of talent. However, Baidu combines companies and colleges, educating talent internally and exporting a lot of talent. It is preparing to set up scholarships for artificial intelligence in colleges and universities.

9

TECHNICAL SINGULARITY:
THE SELF-CHALLENGE OF ARTIFICIAL
INTELLIGENCE

Robots subverting humans is the theme of many science fiction stories, but scientists who want to get things done here and now are still focusing on the current challenges and bottlenecks of artificial intelligence, which reflects the wisdom of human beings.

At present, the outburst of data in the world is almost out of control, and we need a radical innovation to classify and calculate the data. Fundamentally speaking, human beings have not yet fully adapted to the data-based life, as when a human body has not adapted to the rhythm of the assembly line.

Contradiction reigns. We can find some similarities in today's contradictions with those of industrial era.

The relationship between the flying shuttle and spinning jenny is illustrative. In 1733, John Kay invented the flying shuttle. With the prior shuttle, two weavers were needed for large looms. But one weaver could operate the flying shuttle, thus significantly increasing weaving speed. But a problem soon emerged: weaving needs cotton yarn, but the speed of spinning the yarn could not keep up with the weaving demand. The need could only be addressed with more spinning machines and spindles. In 1764, James Hargreaves invented the spinning jenny (with *jenny* being slang for an engine), which doubled the spinning efficiency; the speed of spinning finally caught up with the speed at which the flying shuttle consumed raw material. Several years after the spinning jenny, the spinning mule (a machine to spin cotton) and the self-acting (automatic) mule were invented. By, this time, the speed of the shuttle weaving was not fast enough, so the invention of the power loom was promoted. The two sides, spinning and weaving, inspired each other. At about the same time, the Watt steam engine came to the world, steam force was awakened, and the spinning and weaving sectors were rushing to incorporate this force. At that point, the industrial revolution had been unfolding through countless mechanical advances.

How to Overcome the Malthusian Trap of Data

The relationship between artificial intelligence and data is similar to that between the flying shuttle and spinning jenny. In the past, people conceived a method of machine learning but lacked a sufficient amount of data to verify the learning and practice it. The Internet explosion finally made data available, but dealing with explosive growth of data tests hardware capabilities and computing power.

The Giants' Troubles with Data

The brave attempts of adventurers contributed to the success of big Internet companies such as BAT. These giants share a deep understanding of how to deal with massive data.

During the early stage, Alibaba used the Oracle Database system for data storage. This database architecture of the Internet 1.0 era quickly failed to handle the explosive growth of e-commerce data. Alibaba had to revitalize to build and use its own database.

Before the start of 2013, Jingdong often suffered from server crashes due to the surge in visits. It had to update the back-end architecture and replace .NET technology with Java technology.

Chinese people's deepest anger over data may be the ticketing disaster of the 12306 website a few years ago. Going home for the Spring Festival is a tradition in Chinese people's blood. But being such a populous country, China suffers from digital disaster every year. For the train lines in the physical world, this stresses transportation. Everyone is painfully squeezed in the passenger car without dignity. This situation has been gradually eased by high-speed railways. But the same congestion has shifted to the network. In order to facilitate ticket purchasing, the Ministry of Railways upgraded the informatization of the ticket-purchasing system and launched the 12306 website. However, at that time, nobody expected the data challenge brought by Internet. It was intended to facilitate ticket purchasing, but it was ended up in inconvenience. Hundreds of millions of people searching and buying tickets at the same time quickly made the server crash. People blamed programmers for incompetence and claimed that replacing them with e-commerce engineers could solve this problem. But the real key factor was that processing ability couldn't keep up with data development. Someone specifically compared the e-commerce website with the 12306 website. In a Double Eleven (November 11th) sales promotion, Taobao and other e-commerce websites also took orders from a large number of people, but the companies distributed a great deal of goods with few problems. But the comparison is invalid: tens of thousands or even hundreds of thousands of people rushed to buy little more than a thousand seats for each departure. With every potential purchase, the ticketing system not only analyzed the data of all the stations of the line but also counted, dozens of times, the number of tickets for the line and updated the number of available tickets of all the stations in real time. One ticket influenced the entire line. The amount of data and calculations grew geometrically, and everything had to be done instantly, which is difficult to solve even with more servers, regardless of cost. Such a problem did not exist for large-scale e-commerce, and it was only alleviated after the new computing architecture and methods were explored.

Baidu is the first company among BAT to face big-data impact. Netizens from all over the country send massive amounts of searching data to the Baidu server. The network information that grows day and night also exhausts Baidu content crawlers. Baidu uses presearching and relevant-word searching to alleviate the transient data impact on the server. In the presearching mode, the system automatically searches for and fixes the search results when the number of search requests are low (such as early morning). When the user sends the search request, the system returns the finished result without searching all over again. Relevant-word recommendation uses the system's relatively idle time and clear functional structure to analyze the user data behavior. For example, when the user enters "TPP" (Trans-Pacific Partnership Agreement) in the search input box, a drop-down menu will automatically pop up to provide search options, such as: "TPP means," "TPP's impact on

China," "TPP12 members," "TPP protocol," etc. Of course, the system will also automatically guess that a few users want to enter the phonetic abbreviation of "Tao Piao Piao" (a movie-ticket buying app), which will also be listed in a nonpriority position for users to choose. The arrangement of these options is understandable and meets the basic needs of most people.

At the bottom of the search results page, Baidu also provides a related word search.

In addition, the search engine also lists most frequently searched news related to TPP according to the searching popularity, which is convenient for users to obtain information.

The suggestions are all made by using the statistics of a large number of user searches, which helps to optimize the searching experience, boost search speed, and ease data-processing pressure.

Data can be a source of an infinite variety of fantastic problems. Data is not only composed of homogeneous bits but also related to different kinds of special human-activity scenarios, which puts data processing in a challenging position. But fundamentally, the problem is still the same contradiction between spinning jenny and the flying shuttle—all the progress of the hardware will be immediately consumed by the amount of calculation and data. Although hardware capabilities are growing very rapidly, doubling every eighteen to twenty-four months at the same cost (also known as Moore's Law[12]), data is growing much faster than hardware. Why?

The Malthusian Trap of the Data Century

The Malthusian theory of population is well known; grain production grows arithmetically, while the population grows geometrically. If there is no big breakthrough in output, then the grain-based production materials can only rely on land expansion, while family population growth is exponential in the absence of birth control. As a result, the population quickly reaches the ceiling. After food crises, people were caught in wars, famine, diseases, and other disasters, and the population was greatly reduced. The industrial

---

12 Moore's Law was proposed by Gordon Moore, cofounder of Intel. When the price is constant, the number of components that can be accommodated on the integrated circuit will double every eighteen to twenty-four months, and the performance will double too.

revolution, agricultural science, progress in technology, and population management have alleviated the Malthusian trap. Today, a similar catastrophe has appeared in the virtual world.

Thomas Malthus's law for the big-data world can be described as follows:

- The population grows arithmetically, while the data grows geometrically.
- The amount of data increases linearly, and the amount of calculation grows nonlinearly.

The population of developed countries has increased slowly; some countries have even experienced negative growth. But the data generated by the world is always growing at a fast speed. This is because data is generated from all people and all human activities. As long as we want to record, countless data can be generated "out of nothing."

In the early days, most e-commerce websites only pursued markets and users, with only an emphasis on operations instead of data or data layout. For example, if the user closed an order without purchasing, e-commerce systems did not record this behavior and only deletes it. However, later it was noticed that recording and analyzing the user's failed transaction data is also of value and can be used to summarize user credits, preferences, etc., so then the data was recorded. Every behavior was recorded, and the amount of data began to multiply.

The storage of data has always been a big problem. According to Forrester Research, a well-known market-research firm, one smartphone can generate an average of 1G of data per day. The number of global smartphone users are conservatively estimated about more than two billion, generating more than 2 billion gigabytes (or 2 exabytes) of data per day. If we want to store this amount of data with ordinary 1 TB (with a capacity of 1024G) hard disks, we will need two million hard disks every day and nearly eight hundred million hard disks will be needed every year, which far exceeds the global output.

A more frightening fact is that most data is not generated from human activities. In 2014, Imperva Incapsula, a website security and content-distribution company, released a statistic: 56 percent of page views were contributed by crawler robots. In other words, the main Internet users are no longer human. Most of the click data is generated by machine programs.

Imperva Incapsula's data comes from fifteen billion visits in ninety days from twenty thousand websites around the world with at least ten visits per day.

Nearly half of nonhuman access comes from benign robots, such as content crawlers of search engines, which can index web pages so that people can find the corresponding web content quickly. Baidu and Google use this method to organize information. However, more than half of the page views also come from malicious robots, such as pirate crawlers who steal content, various hacking tools, spam-sending tools, etc., and the proportion is still increasing. In a sense, this is a rich picture of the dark side of the Internet. This is only for web browsing. In the entire human society, with the rapid development of informatization and the Internet of things, all the hardware elements connected to the network are producing data and communicating with each other. The detection chip on the generator set detects the running status and sends the data back to the server. The cameras across the city upload the monitoring data to the command center, and the smart TVs, refrigerators, etc., in our homes are collecting and uploading data. Even if all humans are asleep, the world is still moving forward according to the rhythm of the data ocean.

The amount of data increases linearly, but the amount of calculation increases accordingly with a nonlinear index. Data must be processed to reach its value, but as the amount of data increases, the amount of calculation will increase at a faster rate. For example, the number of squares of Go is only five times that of chess, but the calculation is hundreds of millions of times more complicated than chess. For e-commerce and search engines, if the list of products or search results is sorted, the amount of calculation for this work will rise with a steep curve as the number increases.

Overcome the Trap of Data

To cross the Malthusian trap of the data century, we need to do three things:

- Deal with a large amount of concurrent data in an efficient way
- Store data efficiently and delete unnecessary data
- Mine the accumulated data

"Thinking" Pushes "Physiological" Revolution—Innovation in Hardware Infrastructure

The increase in amount of data and computation necessitates a corresponding change in the entire information infrastructure. All the aforementioned data-processing methods are still developed on the old information infrastructure. But the development of machine intelligence has called for "physiological" changes of the machine brain.

Green Body

The human brain only accounts for about 2 percent of body weight but consumes 20 percent of the body's total energy consumption, nearly 20 percent of the daily total oxygen consumption, and 75 percent of the blood glucose stored in the liver.

The same is true with what is happening in the field of machine intelligence. Data and algorithms are not physical, as opposed to matter and hardware, and are instead analogous to thought. But the operation of this "thinking" requires huge material resources and energy. In those large data centers, in addition to piles of servers, there are power supplies of all the sizes, environmental-control devices, monitoring devices, and various security devices that run around the clock like the brain. The organ itself also consumes a lot of energy.

The Internet provides 24/7 uninterrupted service, and the server consumes a lot of energy. According to statistics, only in 2011, China's data center consumed about 70 billion kilowatts of electricity, accounting for nearly 1.5 percent of the total electricity consumption of the whole nation, and is equivalent to the annual electricity consumption in Tianjin.

In March 2015, the Ministry of Industry and Information Technology, National Government Offices Administration, and National Energy Administration formulated and promulgated the National Green Data Center Pilot Program, which revealed several figures:

> With the rapid development of information technology, the construction of global data centers has obviously speeded up, and there are already more than 3 million data centers in the world now. The electricity consumption accounts for about 1.1 to 1.5 percent of the global amount. The problem of high energy consumption has attracted

the attention of governments. Currently, the average power usage efficiency (PUE, or total energy consumption of data-center equipment) of the United States data center has reached 1.9, and the PUE of the advanced data center is less than 1.2. In recent years, China has built over 400,000 data centers with a rapid development. The annual electricity consumption exceeds about 1.5 percent of the total electricity consumption of the whole nation. The PUE of most data centers is still generally above 2.2, which is a big gap when compared with the international advanced level, and there is huge potential in energy saving. Power-saving, water-saving, low-carbon, and other technological products and advanced management methods are widely used to build a green data center to maximize energy efficiency and minimize environmental impact. The US government has already implemented the Data Center Energy Star, and Federal Data Center Integration Plan, and the European Union also implemented the Code of Conduct of Data Center Energy Efficiency. The International Green Network set the standards for data-center energy efficiency and best practices, which promote the improvement of energy conservation and environmental-protection level of the data center.

What exactly do we do?

Cooling down the equipment room requires constant innovation. Large companies can choose to place data centers in cold regions close to the poles, which made Iceland an important location for major data events in recent years, such as the WikiLeaks event (in 2010, Wikileaks released the US military's classified documents about the war in Afghanistan) and the movie *The Bourne Identity*. Seawater cooling or air cooling are used to save energy and protect the environment.

Yangquan City, Shanxi Province, is an ancient city at the foot of Taihang Mountain with a long history. Li Yuan, Emperor Gaozu of Tang Dynasty, who attacked the central plains, set up a base camp 200 kilometers west of Yangquan. When Liu Cixin worked at the Niangziguan Power Plant, which is about 50 kilometers northeast of Yangquan, he wrote the world-class science fiction novel *Three-Body Problem*, which constructed the "sociology of the universe." Yangquan is a coal region in China, where air pollution is more serious than in Beijing. It has long faced the challenge of industrial upgrading. The tension between history and the future is like the smog that spreads over the mountains and rivers.

In 2015, the Baidu Cloud Computing (Yangquan) Center (hereafter referred to as Yangquan Cloud Computing Center) was put into use. After its completion, the data center storage capacity reached more than 4,000PB (petabytes); the amount of information that can be stored is equivalent to more than two hundred thousand times the total collection of National Library of China. The total number of CPUs in the data center was up to seven hundred thousand, and the total amount of CPU cores exceeded three million. High-performance, low power-consumption servers and a number of other technologies that are applicable for China's environment and regulations were installed to improve the overall energy efficiency of the data center with a PUE of less than 1.3. This means for every 1.3 kilowatts of electricity consumed in the computer room, 1 kilowatt is used for data calculation and 0.3 kilowatts for all other purposes, such as heat dissipation, which is first-class level in Asia in terms of green energy conservation.

Adhering to the spirit of openness, Baidu cooperated with Tencent, Alibaba, China Mobile, China Telecom, and other related industry leaders to jointly establish China's first hardware open source project: Project Scorpio, aiming to create an open technical standard and to develop customized, full-rack server solutions to meet the data center's massive computing and storage needs and effectively reduce data-center procurement and deployment costs.

In September 2014, Project Scorpio upgraded its server technology specification to the 2.0 version, with a more refined definition of space utilization and cooling strategy and a detailed definition of modules, interfaces, and protocols. Based on these standards, a double-digit power-consumption savings was realized through the integrated design of various resources in the Yangquan Cloud Computing Center. Through this plan, we see rapid iteration of the Internet in which products and services can quickly adapt to the changing needs, and constantly launch new versions to meet or lead the needs, always faster than the competitors. The brand-new 3.0 specification emphasizes more on modularity, while the details of the specification are more comprehensive and enforceable.

In May 2015, the solar photovoltaic power-generation project of Yangquan Cloud Computing Center was successfully connected to the power-generation grid. That was the first application of solar photovoltaic technology in a Chinese data center, which reduces 107.76 tons of carbon dioxide emissions per year and saves up to 43 percent of energy.

Computer Architecture Innovation

Although the steps are important, the energy-saving and emission-reducing methods are an external change, such as blowing cold air with an energy-saving air conditioner on someone with a fever; we need internal innovation of computers. Just as the batch processing was the product of the old era, the existing server and data center architecture is also built on the old-world computer technology, which is half old, half new.

The traditional computer core architecture is based on the von Neumann structure: separation of data storage and processing, and linearly distributed computational logic. The computing chip executes the instruction code and stores the result in memory for the next calculation instruction to call it. Such a structure is very clear for humans, but the speed is greatly affected. Moreover, in such a linear flow, the arbitrary instructions executed by the CPU requires the instruction memory, decoder, arithmetic unit, and branch jump processor to work together, with the assignment based on the order in which the instructions are executed. The logic of the control-instruction stream is complicated; it is difficult to have too many independent instruction streams, and the parallel-processing capability is low.

Moore's Law is now out of date. At present, the annual increase of computer memory operation speed is only 9 percent and 6 percent for hard-disk performance. The running speed of computer memory is only a few hundredths of the CPU speed, which is a bottleneck. The pattern of data storage throughput has seriously degraded computer performance.

In the early days, someone had proposed a computer with changed architecture. Taking a personal computer as an example, it is based on general task. Even if a simple task such as typing is performed, the entire computer system is busy in operation, and all other the resources are therefore wasted. Computers that can change the architecture can call different parts of the computer in a controlled manner for tasks with different levels of complexity, without all the resources being called for both big and small tasks. Truly realistic computer innovations have found a way in the development of new technologies.

One direction is leading-edge physics, such as the fascinating quantum computing, which uses the quantum-state superposition effects in quantum physics to create a million times the performance of today's computer chips.

Replacing current transfer data and operations with an optical flow is also a direction to increase speed. Another direction has been learned through the rise of brain science and deep learning: it is hoped that imitating the human brain to develop neurological chips will lead to computer speed being orders of magnitude faster than existing computers.

People are trying different ways. It is an unprecedented step for deep-learning scientists to use GPUs instead of CPU groups for machine-learning technology. The GPU uses SIMD (single instruction, multiple data) to allow multiple execution units to process different data at the same time. Originally used to process image data, it is also particularly well suited for dealing with nonlinear discrete data that deep-learning tasks often encounter. Baidu uses large-scale GPU clusters to optimize its engineering and developed its own GPU server, which greatly improved hardware performance. But the GPU is also built on the von Neumann structure.

FPGA chips are another popular development. It was originally a solution to the application-specific integrated circuit (ASIC). An ASIC is an integrated circuit for a specific user or a particular electronic system. In the past, digital integrated circuits have greatly reduced the cost of electronic products, thanks to their versatility and scale production. But at the same time, the contradiction between general and special use and the disconnection between system design and circuit production arose. The larger the size of an integrated circuit, the harder it is to change the specific requirements when building a system. To solve these problems, there has been an application-specific integrated circuit that allows users to participate in design features, namely FPGA.

The design of the complex parallel circuit was applied to computing chips. The FPGA computing chip is covered with a "logic cell array" and includes three parts: a configurable logic module, an input and output module, and an internal connection. They are independent basic-logic unit modules that implement both combined logic functions and sequential logic functions, defining their respective logic and relationships with each other in a hardware description language. Unlike the von Neumann structure, memory in the structure has two main functions: to preserve the intermediate calculation results and to perform interunit communication. Since the memory is shared, when multiple instructions require memory to be called, access arbitration needs to be called sequentially; the registers and on-chip memory

(BRAM, or block RAM, a fast and small internal memory) in the FPGA have their own control logic, without any unnecessary arbitration and buffering. The connection relationship between each logic unit of the FPGA and the surrounding logic unit is programmable, so it can be determined in advance without communication through shared memory. Parallel computing is the main operation, and at the same time multiple instruction streams and multiple data streams can be processed, which greatly saves computation time. FPGA can also be specially programmed on the hardware for different application scenarios with high flexibility.

Baidu began the layout of FPGA in 2012. It was the first company to introduce FPGA in China and also one of the first companies in the world to use FPGA for clustering. Ya-Qin Zhang said that from the beginning it was the CPU, and then GPU was used. Basically, all artificial intelligence companies use GPUs. But FPGA has its own advantages. One of many is the improved speed and efficiency of the entire architecture. GPU performs better on image and voice data, but FPGA is faster in many types of general-purpose computing. The programmable FPGA allows the architecture to be changed quickly. The FPGA used by Baidu is currently five to six times more efficient than GPU and CPU architectures and can be accelerated directly without changing the existing architecture.

From a computational point of view, network transmission is often considered as the most important bottleneck. Baidu has invested in the most advanced technology for the entire network communication, using 100G RDMA[14] to communicate between GPU and FPGA. So data can be transferred quickly and flawlessly between clusters and databases.

FPGA is equivalent to programming software with hardware, and it is difficult to implement complex algorithms. Currently it works with GPU and CPU architecture.

Since probability calculation is a mathematical method commonly used by big data and artificial intelligence, some inspired people proposed the concept of the probability chip. The probabilistic algorithm is used to replace

---

14 RDMA quickly moves data from a system to remote system memory over a network without affecting the operating system, thus greatly reducing computer processing. It eliminates external-memory copying and text-swapping operations, thereby freeing memory bandwidth and CPU cycles and improving application system performance.

the previous calculus algorithm, which exchanges calculation precision for a great improvement on calculation speed and energy-consumption reduction. It is suitable for such applications that do not pursue extreme precision, such as the Internet of things.

With the rise of deep learning, chip scientists have been greatly inspired. The most cutting-edge chip innovation belongs to artificial neural-network chips, based on the principle of deep learning. Intel, IBM, NVIDIA, and all other major companies have set their own chip-development direction. The deep-learning chips launched by Chinese companies headed by Cambricon Technologies have already led the world for the category.

The artificial neural network is a general term for the computational architecture that mimics the biological neural network. It is interconnected by several artificial neuron nodes, which are connected by synapses. Here, each neuron is actually an excitation function, and synapses are the strong and weak weights that record the connections between neurons.

The neural network is multilayered, and the input of a neuron (function) is determined by using the output of the previous neurons connected to it and the weight of the connected synapses. The so-called training of a neural network is to adjust the output result by inputting a large amount of data and supervision. This process is to continuously adjust the synaptic weight between neurons until the output becomes stable and correct. Subsequently, when new data is entered, the output result can be calculated according to the current synaptic weight, thus realizing neural network's "learning" of existing information. That is to say, the storage and processing in the neural network are integrated, and the intermediate calculation results become the weight of the synapses.

Traditional processors (including x86 and ARM chips) are subject to the von Neumann structure and are inefficient when dealing with deep-learning neural-network tasks. The storage and processing are separated, and its basic operations are arithmetic (addition, subtraction, multiplication, and division) and logical (and or not). The chips often require hundreds or even thousands of instructions to complete the processing of one neuron, and that is why AlphaGo requires so many chips (the distributed versions have 1,202 CPUs and 176 GPUs).

Chips specially designed for deep learning are different. Take the example of DianNaoYu, which is developed by Cambricon. The instruction set directly processes large-scale neurons and synapses. One instruction can complete the processing of a group of neurons and provide a series of specialized support for the data transmission of neurons and synapses on the chip. At the current technical level, the average performance of a single-core processor is more than one hundred times that of the mainstream CPU, with only about 10 percent of area and power consumption, and the overall performance can be improved by three orders of magnitude.

Of course, the neural network chip only has advantages over traditional CPUs on artificial-intelligence tasks and is suitable for image and speech recognition and similar, while traditional chips are better at performing tasks such as running databases, Office, and WeChat—unless such tasks were to undergo a structural revolution.

Neural-Network Evolutionary Philosophy

As the most basic method of artificial intelligence, the combination of deep learning and neural networks will determine the progress of artificial intelligence. This technique, which mimics the mechanism of the human brain, exhibits characteristics similar to biological evolution.

Competition or cooperation? People often struggle with this conceptual problem, but the philosophy of nature is focusing more on cooperation. In recent years, scholars in different fields have proposed the concept of coevolution. The relationship between the *Danaus plexippus* (also known as the monarch butterfly) and the plant milkweed is a typical example.

Milkweed juice is poisonous, and its closed structure makes it difficult to spread pollen through the wind, but the plant can attract butterflies by its nectar and pollinate through them. The monarch butterfly larvae feed on the young stems and leaves of the milkweed, whose toxins can be stored in its body to defend against enemies. If the butterfly larvae eat too many stems and leaves, the milkweed will die, and some of the milkweed will mutate to a more closed structure that can hinder the butterfly from entering. But some monarch butterflies will have enhanced ability to invade the milkweed-variation stamen. So the two become more and more inextricably involved in the encounter, since the monarch butterfly does not eat other plants and milkweed does not welcome other insects; thus, no third party can join their game. Virus and antivirus software, hacking and antihacking procedures are examples of coevolution on the Internet. Machine learning has now been applied to network security, and its efficiency has been greatly improved compared to past firewalls based on set features. Coevolution is not a life-or-death struggle, nor is it a sigh of relief, but an upgrade in methods of confrontation.

Artificial intelligence is also coevolving. In the ever-changing neural networks, this process is vividly revealed. Two of the new neural network ideas are introduced as follows.

Generative Adversarial Networks

Supervised deep learning means that the input data has semantic labels, and the output results are marked by human beings. But many scientists believe that unsupervised learning is the future direction of development, which allows the machine to find the law from the original data. There are already many different approaches with reinforcement learning being one of the directions, and generative adversarial networks are already in use.

Ian Goodfellow, the inventor of generative adversarial networks, is a student of Yoshua Bengio and now works at Apple as the director of machine learning. Yann LeCun, a well-known deep-learning expert, praised generative adversarial networks. This kind of network can well reflect the entanglement and evolvement feature of "evolution."

Generative adversarial networks are derived from the concept of adversarial examples, which was first introduced by Christian Szegedy and others in the paper published by ICLR2014 (International Conference on Learning Representations). Subtle interference is deliberately added to the input data set to form an input sample, resulting in erroneous output of the deep neural network. This error is obvious in the human's eyes, but the machine can repeatedly fall into the trap.

Ian Goodfellow, Jonathon Shlens, and Christian Szegedy give a typical example in the paper "Explaining and Harnessing Adversarial Examples."

People add tiny interference to a picture of a panda. Modifications are performed in 32-bit floating-point values without affecting the 8-bit representation of the image.

The human eye cannot see the difference at all, but the neural network surprisingly judges the picture as a gibbon with a 99.3 percent confidence level. Because the confrontational samples lead to recognition errors, some people regard them as deep learning's deep flaws. But Zachary Chase Lipton, formerly from the University of California, San Diego, and currently with Carnegie Mellon University, published an article at KDnuggets (a big-data media site in the United States) with the title *(Deep Learning's Deep Flaws)'s Deep Flaws*.[15] This article argues that the vulnerability of deep learning to adversarial samples is not unique to deep learning and that it prevails in many machine-learning models. Further research on algorithms that resist adversarial samples will facilitate the advancement of the entire machine-learning field.

Scientists have grasped the fragility of "evolution" and see nature as making the best from a mistake, regarding confrontation as a training method that turns all the obstacles into motivation to advance through difficulties. The evolution of nature itself is also highly fragile, and countless biological "programs" are eliminated from nature because they are "faulty." Error is the ultimate tool of evolution. Wisdom is rising with difficulty in the process of endless birth and death.

The generative adversarial network is an ability of neural network specially designed by humans to actively generate interference data to train the network. Simply, the generative adversarial network consists of two parts: one is the generator and the other is the discriminator. The generator is like a profiteer who sells deceptive fake goods, and the discriminator is like a superb buyer who needs to identify the authenticity of the goods.

The job of the profiteer is to find ways to deceive the buyer (generating confrontational samples), while the buyer learns through experiences and reduces the probability of being cheated. Both profiteer and buyer are constantly striving to achieve their own goal, while at the same time they are the pursuing advancement under the supervision of each other, like the Blue Army and Red Army confronting fiercely in military exercises, thus strengthening the fighting skills of both sides, without smoke.

The generative adversarial network and the response to it demonstrate coevolution. It is a profound philosophy of evolution, an entanglement instead of a war, which maintains a precarious balance.

Do we want the mature buyer or the superb profiteer? The answer is both; they are inevitable elements of coevolution.

---

15 www.kdnuggets.com/2015/01/deep-learning-flaws-universal-machine-learning.html

What is the use of the profiteer model? In many cases, we are lacking data, which can be complemented by generating models. Making unsupervised samples produces effects similar to supervised learning.

Wei Li and Roderich Groß from the University of Sheffield, UK, and Melvin Gauci from Harvard University, US, developed a new Turing learning method for studying group behavior, based on the generative adversarial network.[16] A group of fish is mixed with some fake fish that mimic real fish movement. How does one judge the fidelity of imitation behavior? It is difficult to distinguish between movement behavior by traditional feature induction methods, and the motion characteristics of the same group of fish are not necessarily similar each time. The team decided that to let the machine automatically build a group model by emulating both sets, allowing the machine to infer the behavior of natural objects and imitations. This deep learning simultaneously optimizes two groups of computer programs, one representing the behavior of the model and the other representing the classifier. The model can mimic the behavior of supervised learning as well as the behavior between the system and other models.

To be more specific, they established three groups of robots, the first being the imitated objects, which perform complex movements according to prespecified rules; the second is imitators, which mixed into the first group, trying to learn and imitate the first group's behavior and to deceive the discriminator; the third is discriminators, whose task is to distinguish the imitator from the imitated in the group movement. As the discriminator became increasingly discerning, the imitation behavior became confusing. Therefore, we can use the trained imitators to build a realistic multiagent model to simulate the group of imitators. This model can be used to study collective movement behavior. For example, a model can be trained according to the crowd movements at popular holiday spots, recorded by the camera to improve the prediction of crowd-movement trends and to issue an early warning of possible congestion and stampede.

The evolutionary iteration of machines is a zillion times faster than nature. In this kind of adversarial generation, the logic the machine acquired has gone far beyond human understanding and may become a "black box." It

---

16 www.arxiv.org/pdf/1603.04904v2.pdf

is a big challenge to choose between the black box and white box and to avoid the incomprehensible danger of the black box.

Dual Network

The dual network seems to be a mirror image of the generative adversarial network.

At present, most of the neural-network training relies on tagged data—that is, supervised learning. But labeling data is an onerous task. According to reports, Google's Open Images data set, Google's open-source image database, contains about nine million images; the YouTube-8M data set contains eight million segments of marked video; and ImageNet, as the earliest image data set, currently more than fourteen million classified images. All three data sets cost fifty thousand employees through Amazon Mechanical Turk, an Amazon labor-outsourcing platform, and required two years to complete most of the well-marked data.

To enable the machine work in the absence of labeled data is the future direction of artificial intelligence. In 2016, Dr. Tao Qin and others from Microsoft Research Asia presented a new machine-learning paradigm, dual learning, in a paper submitted to NIPS (Conference on Neural Information Processing Systems) 2016. The general idea is that many applications of artificial intelligence involve two dual tasks. For example, translation from Chinese to English and translation from English to Chinese are dual, speech recognition and speech synthesis are dual in speech processing, image-generated text and text-based imaging are dual in image understanding, answering and generation of questions in the question-and-answer system are dual, and so is the searching of related web pages by keywords and the generation of keywords for pages. These dual artificial-intelligence tasks can form a closed loop and make it possible to learn from unlabeled data. The key of dual learning is that the model of the second task can provide feedback to the given original-task model; similarly, the model of the original task can also provide feedback to the model of the second task. Thus, the dual tasks can provide feedback to each other, as well as learn from and improve each other.

The use of such a subtle strategy by dual networks greatly reduces the reliance on annotation data from which we can once again find some evolutionary philosophy: evolution is a cyclic process of response and receipt, from A to B and from B to A. They are mirror images of each other, but the mirrors are not clear. They each have half of the secret, without arbitration, but they move forward in mutual guess and reference.

Deep Learning's New Frontier

The foregoing two neural-network methods are only typical manifestations of constantly emerging new methods. In addition to deep neural-network methods, scientists are actively exploring other paths. Professor Zhou Zhihua, a famous machine-learning expert at Nanjing University, presented a creative algorithm with coauthor Feng Ji in a paper published on February 28, 2017, which can be called the "gcForest" algorithm. As the name hints, this algorithm uses the traditional decision-tree algorithm, but emphasizes the tree hierarchy, as opposed to deep learning, which basically emphasizes the number of layers of the neural network. The multilevel decision trees form a "forest." Through some sophisticated algorithm settings, in the case of small data size and computing resources and in the application of image, sound, emotion recognition, etc., its results are at least equal to that of the neural network. This new method is insensitive to parameter settings, and the logic-based tree approach makes it easier to theoretically analyze than deep neural networks, thus avoiding the difficulty of human understanding of the black box problem of the machine's specific operational logic.

Table 9-1: Accurate Comparison of Face Recognition

	One image	Five images	Nine images
Multi-granularity cascaded forest	63.06%	94.25%	98.30%
Random forest	61.70%	91.20%	97.00%
Deep neural network (convolutional neural network)	3.30%	86.50%	92.50%
Support vector machine based on RBF kernel function	57.90%	78.95%	82.50%
K-nearest neighbor algorithm	19.40%	77.60%	90.50%

Source: https://arxiv.org/pdf/1702.08835.pdf

Table 9-2: Comparison of Testing Accuracy in GTZAN Database

Multi-granularity cascaded forest	65.67%
Convolutional neural network	59.20%
Multiple layer neural network	58.00%
Random forest	50.33%
Logistic regression	50.00%
Support vector machine based on RBF kernel function	18.33%

Source: https://arxiv.org/pdf/1702.08835.pdf

According to the think tank AI Era, Professor Zhou Zhihua believes that the methodological significance of deep forest is to explore the possibility of algorithms from outside the deep neural network. The effective operation of deep neural networks requires strong power for data and computing, and deep forests are likely to offer new options. Of course, deep forests still draw key ideas from deep neural networks, such as the ability to extract features and build models. Therefore, it is still a novel branch of deep learning. Chinese scientists have many world-leading achievements in artificial-intelligence research. We believe that self-confidence and an open mind will be an important driving force for scientific progress. Today, major technology companies for artificial intelligence advocate sharing algorithmic code; Google's TensorFlow deep-learning open-source platform is the most influential example. However, many deep-learning scientists believe that from an economic point of view parallel competition between more deep-learning code platforms will be conducive to prosperity and a counterbalance to monopoly. In addition to the deep-learning open-source platforms such as Caffe and MXNet from other companies, Baidu opened PaddlePaddle, a deep-learning open-source platform, in September 2016. With the new architecture, it has good support for serial input, sparse input, and large-scale data model training. It supports GPU computing, data parallelism and model parallelism, and training deep-learning models with a small amount of code, which greatly reduces the cost of deep-learning technology. A diverse shared platform enables machine learners to train and create applications from different perspectives, a type of biodiversity to contribute toward the advancement of artificial intelligence.

Even if only in the remote future, artificial intelligence can really be powerful enough to rule the world. All challenges thus far have been related to human wisdom. But the flash of wisdom from artificial-intelligence scientists illuminates the direction for the latecomers. Even non-artificial-intelligence practitioners may get a great deal of strategy and inspiration.

At the beginning of 2017, Master, a version of AlphaGo, swept the top players from China and Korea. For a time, people have been divided into different categories: pessimists, adventurers, calm people, outrageous people. We hope that more people will be in the silently learning category.

10

MEETING YOU IN THE INTELLIGENT AGE

On January 6, 2017, Duer the robot stepped on the stage of the *Super Brain* TV show to challenge humans for the first time. On that day, it received mostly admiration from the audience. It was able to identify a certain adult girl from the crowd and her twin sister, based on a picture of the girl in her childhood.

Duer is very cute, but it makes human players and TV audiences think, *When machine intelligence really surpasses humanity, what will our life be like?*

On careful examination, we will find that artificial intelligence has long entered into every aspect of human life. It occasionally plays chess or some other intellectual competition with us. More often, it is a babysitter, teacher, housekeeper, assistant, driver, or doctor. Let's experience an "inception" and meet you in the future world:

On a calm afternoon, you just finished 3D-projection work on conference at home. Walking out from the room, you see the new child-companion robot teaching your son to learn the math multiplication table. Your wife comes in with fresh fruit grown and picked from the smart orchard. Your son runs and hugs his mother and whoops over an animated film that has just been released. Why is the little guy so well informed? You checked the smart TV and found that the TV that had been switched to the Son mode sent him the news of this film.

You promise your son to watch the film. But you didn't order a home-theater version as usual because the name *Goodbye, Castle in the Sky* evokes memories of your childhood about a Japanese film that discusses the relationship between people and machines, and this one is a new entry in the series. You decide to bring the family to the cinema.

In the era of robots everywhere, you feel that places like cinemas are necessary for children because of the contact with others. Interestingly, although home theaters are so popular nowadays, the number of cinemas in the city has not yet decreased; instead there are more moviegoers. Perhaps in the age of artificial intelligence, human nostalgia is awakened. The cinema where you used to go when you were a kid is at the other end of the city, but you want to go there. So, you use your voice to wake up the ubiquitous Duer system in your home. Duer tells you that the cinema is still there, picks the right time, plans the route, and orders the unmanned vehicle.

The unmanned vehicle already knows the destination of your trip as you get in. The car screen is playing a clip of *Castle in the Sky*[17] instead of the director interview about this new film, which would obviously have been for your son's sake. You wake up the screen by voice and connect it to your phone to deal with some emails. At that point, there is a video call from the family of your wife's good friend. The artificial-intelligence housekeeper at your home told your wife which friends were interested in the film, so she sent a message about going to the cinema before leaving the house. The wife's friend responded to say that her family arrived and bought the tickets.

The cinema has well-preserved historical décor. You no longer need 3D glasses for this type of movie. The mature naked-eye 3D technology makes the viewing vivid and comfortable. It is said that the sensors in the cinema will record the emotional response of each in the audience during the movie, which will be analyzed with the plot to generate a consumer report and sent to the creative team and distributor as feedback.

After the movie, the smart assistants of the two families have each recommended nearby restaurants for dinner, and finally humans decide which one to go to. The dinner time is more than happy. On the way home, the intelligent audio of the unmanned vehicle invites your family to talk about the movie. Your son agreed gladly. AI asks if he likes this movie, which character he likes the most, and how many points you would rate it. You know this is connected to an intelligent film-evaluation system; the content of the conversation will be sorted and refined for the reference of the producer. You can also see the movie-review conversations that others want to share. You want AI to have a more advanced, purer dialogue, such as asking the child what he wants to do with the robot. So, you ask, "Can you help me to rule the world?" AI replies, "I came to serve the world, not to rule it."

Artificial intelligence makes the world better, not worse. Most people pay more attention to health, life, and learning because of their intelligent assistants. Your child will become a new human in the intelligent age.

---

17 A film directed by Hayao Miyazaki, a famous Japanese animation director, and released in 1986, about the relationship between man and machine.

Basic Necessities of Life in the Intelligent Age

At the beginning of 2016, Mark Zuckerberg announced an intention to create an AI housekeeper. By the end of the year, he uploaded a video on his Facebook account to showcase the results.

In the short video, Zuckerberg talks to the intelligent system that he spent more than one hundred hours personally developing. In the video, the smart housekeeper only completes functions like ordering songs, adjusting the lighting, and recognizing visitors' faces. Zuckerberg named the system "Jarvis," in honor of the famous character of the powerful intelligent system in *Iron Man*.

At the grand CES show in 2018, the smart housekeeper was an intelligent video speaker called Duer Home, which won praises from many in the global technology media for its user experience, shape design, and built-in conversational AI operating system DuerOS (known as the Chinese version of Alexa).

This smart video speaker, which was officially launched in China on March 26, 2018, is equipped with the latest Baidu DuerOS conversational AI operating system, which combines a six-microphone far-field voice, high-quality speakers, touch screen, and camera. It can listen, see, talk, and think. In addition, Duer Home has stored thirty million short videos, fourteen million encyclopedias, five hundred thousand children's stories, one million comic dialogues, one million recipes, and hundreds of millions of maternal and child information.

Duer Home can support a variety of hardware devices, such as mobile phones, TVs, speakers, cars, and robots, and can also support third-party developers. Users can wake it up by saying "Duer, Duer," and then let it play music or broadcast news, search for pictures or information, set alarms, order food deliveries, or chat. User can also initiate multiparty video calls, voice messaging, voice photo display, and video display. DuerOS is equipped with Baidu speech recognition and natural-language-processing technology, and it can keep learning and evolving through the cloud brain. These technologies are not very new, but they have realized the functions that once existed only in ordinary people's imagination.

Suppose you have your own Duer Home. What do you say to it first? Maybe something like, "Turn on the TV" and "Turn on the kitchen light." Deep-learning-based AI can gradually understand the deep relevance of language through training. In the future, smart home appliances will understand the complex needs of the owner through a simple sentence. For example, when you say, "I want to go to bed," it will lock the door; turn off the lights, except in the bedroom; turn on the bedroom air conditioner; etc. When your child gets up in the middle of the night, half asleep, and wants to go to the bathroom, he will only need to say, "Pee." Then the night-lights in the bedroom and bathroom will automatically be turned on, and the toilet will also start the automatic flush mode, without waking up the tired parents.

The smart home system not only understands the family's living habits through "learning" but also becomes a childcare expert, work assistant, and elderly care professional through big data and deep learning of millions of families. For example, it will advise about the baby's sleep time, remind about flu-prevention techniques according to any current flu outbreak, and notify the elderly where they can exercise square dance. The humanized smart home not only makes future life more comfortable but also connects people with the world.

In addition to the smart-home system, the issue most closely related to people is about eating. When the vegetable delicacies are still growing in the soil, the improvement of food by artificial intelligence has started from the "root."

In 2016, *Wired* published a long article detailing how AI can transform modern agriculture: with image-recognition technology, farmers will be able to identify crop diseases in time to avoid mass crop loss. Agricultural robots can completely reform agricultural labor. An agricultural robot called LettuceBot looks like a tractor, but it can scan over five thousand seedlings per minute, automatically distinguishing weeds and remove them, ultimately reducing 90 percent of herbicides. To teach about weather issues, scientists have included satellite images into the training of deep learning. In the future, a farmer will open the app every morning and check the detailed climate conditions on his land, accurate to several kilometers.

An intelligent unmanned dairy farm has been built in the Netherlands, and the entire pasture is operated by an AI system for dispensing feed or milking.

All of these will make future agricultural products cleaner and more productive. Soon, only the goods with the "Smart Farm Produce" label on the package in the supermarket will be recognized as safe vegetables or meat.

With good ingredients in hand, we need good cooking. We always said that artificial intelligence will replace human mechanical work, but cooking is not a simple mechanical production. The completely different cuisines of all corners of the country and world, as well as everyone's taste, makes cooking a matter of sentiment.

The main purpose of the technology has never been to replace but to support. For many people, the cooking process itself is a pleasure, which is strengthened by AI. To be sure, the AI system would help the user to perform some simple operations, such as cracking eggs, adding water, and adding flour, so that the human chefs can concentrate their energy and time on the taste. But it could also learn the user's operating habits and taste characteristics through deep-learning techniques to cook dishes that taste like "grandma's dish." Through data sharing, it would send delicious recipes to more people's cooking system. At that time, the cooking system will recommend and share dishes based on the user tastes, just like today's interest-based news clients recommend information.

The fun of eating out will also be changed. It's now common to choose a restaurant and order dishes using a smartphone. You can take a photo of a coveted but unfamiliar dish. Duer will help you to identify the name, taste, ingredients, and nutrients. Some users who are keen on food pictures with their circle of friends may lose a little fun in the future because of less mystery in the dishes.

In the field of entertainment, all the movie and game companies who can understand user psychology can surely sense the business opportunities brought by AI. Gamers will soon get along with AI. People have been fighting AI in games, but all the "artificial intelligence" there is still typical symbolic intelligence that operates according to the symbol program set by human beings. Flexible players soon find the loopholes of such "artificial intelligence" and crush it.

Players often say, "I will beat you like I beat the computer" when they find their competitor a terrible player. But this comparison may become a compliment in the future. Google had trained artificial intelligence to operate StarCraft II and recently challenged professional e-sports players. The AI agents won ten consecutive matches before losing to a human in the final match. For the games, AI was restricted by similar conditions that bother a human, so they acted like human players, to the point that discerning whether the competitor was human or artificial was difficult. The AI player could not click more times per minute than a human, for example, although it could see the whole map at once rather than navigating it like humans.

Mature AI will alert you when you indulge in the game so much that lack-of-exercise health indicators have kicked in. The new life of the intelligent era should not be around the computer, and artificial intelligence should not keep users completely at home. In the future, people will only need to carry their mobile phone when they go out; then, AI will entertain them, and those who had not been interested in traveling will have new fun.

Duer's help with booking, reservation, and route planning is, in a sense, metaphysical. On the spiritual level, artificial intelligence rewrites the meaning of travel.

The machine can even understand our mood. The AR (augmented reality) technology developed by Baidu Deep Learning Lab allows the mobile phone to tell you about the flower nomenclature, the name of birds, the legend of the mountain, and the colloquialism of the ancients. In the face of fleeting beauty, it can help you to capture the purest sunsets and evanescent waves. Combined with the map tool, it can also send other people's tips and experiences to you each time you arrive at a new place.

The ubiquitous AI interpreter allows you to talk to foreigners. When talking to each other in a strange language, AI seems to be simultaneously whispering.

Apart from different languages, AI can also connect light and darkness. A disability robot implanted in a mobile phone or wearable device can use visual and speech technology to help the blind to judge the value of currency and remind them about the surrounding environment. We can already envision robot guide dogs in the future.

We can even imagine the next-generation map plan, which uses drones to collect more stereoscopic map information in the traditional way and utilizes 3D-reconstruction technology to restore the real world in all directions. AR navigation gives us a pair of "perspective eyes," and the next stop, whether a store, an airport or a hospital, will be unobstructed. Image-recognition technology keeps the map information updated all the time, and the road ahead and the scene are predicted before you take the next step. Virtual reality allows humans not only to go through the barrier of language and light but also to walk through dreams and reality like *The Matrix* movie.

The current artificial-intelligence service is far from perfect, but the big goal is undeniable, with different roads leading to it. To have the initiative to live in the future, we must prepare for these advances. Some people worry that future social artificial intelligence will only belong to a few knowledgeable people, but we hope that more people will be able to share the dividends and amenities of the intelligent society.

Don't Lose at the Starting Line of AI

In April 2016, Chinese writer Hao Jingfang won the Hugo Award for her science fiction novelette "Folding Beijing." This is the second time that a Chinese writer received an "Oscar" Award in the sci-fi world, after Liu Cixin's *Three Body Problem* (Hugo Award for Best Novel).

In the face of AI where "the future has come," Hao has her own point of view:

> The future human society is more clearly divided; standardized production is done by AI, and human beings are responsible for creative and emotional work, and the organizations of future society will be more flexible. The most crucial point is that in the society dominated by AI, the current education model can no longer adapt to change. It is difficult for us to face the future era of AI without reform.

This speech was made by Hao at the Sina TECH Top Awards in 2016. Young parents may miss an opportunity if they just expect the children to have a life trajectory like hers but ignore her advice on education.

Chinese parents are often concerned about their children's education. In the eyes of many parents, a good school means a good education. But do parents who focus on teachers and enrollment rate realize that what qualifies as a good school will change in the age of intelligence?

AI is connected to various aspects of teaching, learning, and management in school. The Internet and AR technologies allow teachers and students to interact across time and space. The teaching scene goes beyond the traditional classroom. In virtual space and online education, students are more equal and time scheduling becomes more flexible. Students can better prepare before class and study in more convenient grouping, while teachers can also adjust the progress and teach students in accordance with their aptitude.

The intelligent education system can automatically record each student's wrong answers or slow progress and then match the students with special counseling resources. The lessons that students have learned, the homework they have done, and the materials they have read are not only stored in the profile space but also become rich labels that describe their learning curve and style. The machine secretary can accurately send teaching suggestions and resources to students and teachers according to the label and change the past cramming teaching methods. All students' learning records and feedback will be integrated through artificial intelligence for referencing, optimizing, aggregating, and distributing, which will inspire each other while personalizing education, improving the overall level, and completely upgrading the meaning of "two-way education."

Teachers are the busiest people in the school. Intelligent systems will greatly reduce their burden and free them from the mechanical work for personalized and innovative teaching. For example, an artificial-intelligence system based on natural-language-processing technology can be trained by massive data to correct homework in seconds for both English and Chinese compositions. Artificial-intelligence systems based on speech-recognition technology can lead students to read English and correct their oral language. In addition to the homework correction, marking papers has also been put into use.

Having the AI system correct homework can have a profound educational impact. Educational research has long shown that human memory follows a law, by which we forget much at the beginning period of learning before tending to stabilize. AI systems can help teachers to give feedback to students faster, so that students do not limp away because of the delay.

In addition to having less of a burden, teachers can understand the overall situation through the operational data, grasp the direction, and truly become an educational conductor and artist.

Robots are already trying to take the college-entrance examination. This does not mean that they will go to college; instead, they will undergo training. A robot that can perfectly cope with the exam will in turn train students. Now, Duer can help with filling in a college application. Many ordinary families often do not know which school to choose because they lack an understanding of colleges and universities. Duer can not only check the score but also provide suggestions based on the big-data status and student-evaluation submissions of previous years.

AI applications are experiencing an unbalanced development in the world; only a few schools boldly start AI-aided teaching, with a high developing speed. According to statistics, nearly half of the national key middle schools have adopted intelligent education systems that work with more than half of active students, and the teaching efficiency has increased by more than 20 percent. In the future, all schools may need an artificial-intelligence "teacher," just as all companies need a CAO. Laggards will not be able to pioneer in the intelligent age. Therefore, parents who have tried their best to choose a school may also look forward to the future and make choices according to artificial intelligence.

There are already many educational robots for children in the market. Some can tell stories for children, some talk with children in English, some can take a photo for the children from a perfect angle and send these photos to their families, and some can even guide children to read paper books. They vary in shape and size, as well as the smart-flow technologies that are connected.

Behind choosing schools and picking toys is the initiative to change the way parents think in the new era.

Imagine how the generation gap between us and our children will change. Will the digital gap be similar to the gap between the parents and the post-80s and post-90s children who grew up in an era of the Internet?

Children can accept artificial intelligence faster than we think. More and more primary schools started programming courses, and more and more children have been exposed to computers and smartphones at an early age. Games and the Internet are also full of various smart technologies.

American TV has conducted an interview with children age ten and older and found that many American children do not know how to use Windows; instead, they are very skilled in operating smartphones. Microsoft now attaches great importance to artificial intelligence and develops AI technology as an important strategy for the whole company. It invests huge resources to develop Microsoft Xiaoice. The sense of crisis brought about by the transformation of a new generation of users is one of the important reasons.

Baidu was also influenced by the changes of the new generation. Jing Kun, vice president of Baidu and general manager of Smart Life Group (SLG), describes the new generation that grew up in the AI environment:

> In the research and development of AI and Duer, Baidu found itself not teaching users how to use it. Instead, it follows the user's need to make products. And children are the most unhesitating and unconstrained in expressing their needs. Children's demands are unrelenting, but they still have great mental and physical deficiencies. Most of the time, they need adults to help them to accomplish their goals, so the needs are always limited. AI can understand their needs by only one sentence or one simple press, which lowers the threshold to achieve their goals and realize their needs. So, when they discovered a voice dialogue system that never gets bored at questions, they naturally learned to use this new way to get information.

Artificial intelligence, like Prometheus's gift, may be the fire of enlightenment or the harmful flame. Only complete domination can help it to play a positive role.

We want to build a high-speed train for new life. We must work hard to carry more people, unlike the ark in the movie *2012*, which is only built for a few people. We hope that when artificial intelligence becomes super-dimensionally advanced in technology the digital gap will also be reduced. Artificial intelligence must be open and rooted in cyberspace, breaking the gap between space and time, so that everyone can access the intelligent flow if he or she works hard.

There are more opportunities and challenges in the smart society waiting for our children. When they graduate from college to look for jobs, what will they face? When robots begin to take over the simple mechanical work of humans on a large scale, how can they harness a new future?

Working Is Beautiful, Especially in the Smart Age

Getting out of the ivory tower for job hunting is another new life threshold. What graduates are most afraid of is not low wages, hard work, or even failing to find a job. It is discovering that what they have learned is useless. Artificial intelligence brings new tools, just as the industrial revolution brought textile machines. The intelligent age calls for new skills and new thinking. In the face of the future, many seniors might wonder, *Can I adapt to the work of the intelligent age?*

The anxiety of making a living hangs around the young people of every generation. There was a popular game on the Internet in China called Life in Beijing, in which the player is a young man trying to make a life in Beijing, with a goal to earn enough money to repay his debts within forty days. This 70KB small game with a simple interface was very popular at that time, because it was full of the element of being adrift in Beijing. The game producer himself was a complete Beijing drifter. In 1992, he came to Beijing with only 200 RMB (renminbi, the official currency of exchange; 200 RMB equals roughly $28.24) to study and work. The game is full of entertaining but true characteristics of Beijing. In the era with no smog but only sand-storms, it has triggered the resonance of countless young people who are struggling in big cities.

In recent days, the producer of Life in Beijing showed up on the Internet. He told many old players of the game, who are concerned about his life, that when he was developing Life in Beijing, he was working on AI research at Beijing University of Posts and Telecommunications. Later, he left the research field and concentrated only on developing games. Recently, he returned to the AI field to study the application of deep reinforcement learning, which serves the vertical field of the game indus-try. In an interview several years ago, he called this little game a "literary work." He said, "I know about Shanghai in the 1920s and 1930s through *Midnight Shanghai*, and Latin America through *One Hundred Years of Solitude*. After ten years, I will be more than happy if someone can vividly understand some of Beijing's life scenes and certain aspects in 2001 through Life in Beijing."

The magical achievement is the "imagical" hard work and opportunity. The year 2001 was the winter of the Internet in China. Various bubbles burst, and many companies closed. Most of the people who went through it that time were later rewarded. Today's leaders on the front line in the emerg-ing fields of mobile Internet, big data, and artificial intelligence may once have been exhausted in a rental house near Fifth Ring Road after working hard on a weekend in 2001, turned on their desktop, and played Life in Beijing before going to bed with a smile.

Internet+ makes all industries inseparable from the Internet. In the era of artificial intelligence, AI has been integrated into all kinds of work. The sense of big data and intelligence determines the level of future workers.

Many mechanical labor positions will be replaced with artificial intelligence. Assembly-line workers, stenographers, taxi drivers, etc., will quickly disappear in the future, just like the decline of paper media in the Internet age, despite of having long history of hundreds of years. Humans should be engaged in more creative work, but first they need to master the tools of creation. Otherwise, even a seemingly high-end white-collar job will be machine-like.

When famous publications such as *Forbes* began to use robots to write their news articles in thirty seconds, journalists who used to be the crownless kings began to feel as if they were sitting on a spiked rug. But there is no need to be too nervous. In addition to working hard to improve the writing skills, they can also use smart tools such as speech recognition, shorthand software, smart editing video tools, etc., which will make journalists and editors more powerful.

The use of data-analysis software will allow reporters to find more news clues. For example, Baidu Index and similar products, rooted in big data of search engines, can provide users with various information dimensions such as search heat, trend change, and interest map of users, with which media practitioners can judge the trend of news topics. Fans can learn about the changes of the popularity of their idols. Shops can observe the user's target and demand trends of products.

Relevant thinking is fully reflected here. For example, in the week June 16–20, 2014, the phrase *ask for leave* was the most frequently searched in Baidu Index, and also the word *beer* was popular. Keen readers can already get the answer from the relationship of the two: the World Cup in Brazil started on June 13, followed by the most intensive match week. What can journalists, brands, and shops do with such data insight? Those good at using these tools will become the best in their industry. In the process of using data tools, we are also being trained to make ourselves more capable of discovering and creating value in the age of intelligence.

Translation and interpreting will also face new challenges. The interpreter who always appears in photos with important leaders will probably be accompanied by AI. The development of AI language-processing tools should focus on new features for high-end customers in addition to basic translation for ordinary people. High-end translations require more than basic machine translation; they use deep-learning techniques to help in grasping the language features of the clients, such as leaders, and provide specific functional services, such as how to translate ancient Chinese poetry into English.

When Baidu Brain conceives verses, when the neural network imitates Bach's talent to compose music, when the Japanese-novel awards officially announce literary works created by artificial intelligence into the candidates, everyone begins to realize the field of culture and art as the holy place of human soul is no longer mysterious for artificial intelligence. The neural network is much like the human brain, and there are countless hidden neural pathways. Li Bai wrote poetry while drunk, because the state of drunkenness activated some of the sleeping nerve pathways. In the future, humans will use the neural network to develop inspiration without having to be drunk.

Professionals such as lawyers and financial experts will also face changes. The world's first artificial-intelligence lawyer, Ross, developed by IBM, worked at Baker & Hostetler in New York in 2016 to help in dealing with corporate bankruptcy. This legal omnipotence can truly put its heart and soul into the job. It is not as expensive as a human lawyer. Why don't ordinary lawyers or judges try this? In December 2016, the Beijing Municipal High People's Court launched an AI system called "Judge Wisdom," which not only does auxiliary work of collecting information and case portrayal but also helps to analyze the case, collects multiple types of information, and identifies the relevant factors that can affect the conviction and sentence and the reasons for appeal, to help the judge make the preliminary judgment, which is quite close to adjudication. Law firms and courts that use AI assistants could handle cases more efficiently, creating higher returns and social benefits.

The financial castle is built up by digital bricks, and artificial intelligence is the cement that penetrates everywhere.

The famous Michael Bloomberg is the Wall Street wolf that is armed with data. The billionaire entrepreneur who was once the mayor of New York City had entered Wall Street through financial data-processing tools decades ago. Before he launched his bond-trading products, traders basically checked the price through Reuters and then calculated with pencil and paper. Bloomberg's products provided real-time data to users, computing power, financial analysis, etc., and quickly dominated all the bond exchanges. Bloomberg first sold his products to Merrill Lynch in 1982, the same year Microsoft just launched the precursor to Excel, which was called Multiplan.

Bloomberg's example may be a bit distant. But in Zhongguancun, Wudaokou, Xierqi, Zhangjiang Town, and Huaqiang North, countless ambitious young people are trying to use data-intelligence tools to change their lives. AI has nothing to do with class origin; it only cares about ability. Personnel administrators can use data tools to study the distribution of company personnel and propose optimization opinions. Coffee shops can order coffee for users through the Smart Dialogue app and quickly know about the users' taste. Former game master players can become live broadcast commentators because they are more likely to discover in which aspects artificial intelligence is not "human" enough.

Although the trend of AI will eliminate some occupations, it will also spawn new opportunities. The car replaced coachmen but also created the position of driver. The popularity of artificial intelligence will also bring new opportunities. Andrew Ng gave an example: unmanned scenarios will become popular, and someone should be responsible for the drone-traffic management, which does not exist yet. Who will do this? He suggested that people put more energy into education and cultivating talent.

What really determines the future of our work is whether we are professional enough in the smart age. As smart tools become more omnipotent, the flexibility to use it forms a gap. A person who knows nothing about programming, data analysis, or machine learning is like someone who can't use wrenches in the industrial age.

Almost everyone owns a smartphone today, but most people use it only to play games, watch movies, and browse a friend's update, without realizing that there is an entrance to the future. Many of the world's most advanced technologies in the field of AI have been integrated into products such as Duer. The seemingly uneventful functions such as translation and speech recognition have made us accessible to cutting-edge of artificial intelligence technology.

Working Is Beautiful is the name of the novel of late female writer Chen Xuezhao. She was one of the pioneers of the New Culture Movement, which sprang out of student demonstrations in Beijing on May 4, 1919. Today, artificial intelligence brings another new culture movement. In the smart age, working is still beautiful and requires wisdom as well as the mutual encouragement and progress from human beings and machines.

Song of Life: How Artificial Intelligence Re-creates Medical Care

Thinking, struggling, enjoying, running—human beings fight hard in the world, whether happy or distressed or eager for health, longevity, or the opportunity to see the future. Long ago, Emperor Qin Shihuang sent people to the Eastern Land for longevity medicine, while today's medical technology has brought longevity to ordinary people.

The development of science and technology has brought new imagination. Consider Baymax the medical robot in *Big Hero 6* and the human-body-transformed weapon in *Ghost in the Shell*, which have become classic and popular characters on the screen. Although human civilization has developed by leaps and bounds after the industrial revolution, the human body is still fragile under the protection of modern medicine. Inside the body is a world we have not completely explored. Where will high technology lead the human body? This is not just an interesting topic to doctors.

Advances in all fields of science seem to be significantly connected with the medical industry. Physicists' research in optics and atomism have brought microscopes and X-rays. The health testing room for astronauts in the spacecraft turned into an intensive-care unit later. Biology makes animal experiments one of the most important medical-testing methods for human beings, and chemistry is a discipline that is inextricably linked to drugs.

So what role does artificial intelligence play in the medical field? The answer is probably not one single role, but it will change the entire medical industry. Artificial intelligence will have a comprehensive impact on medical care, both elementary and profound, partly and entirely.

Cutting-edge medicine is the most eye-catching: *Nature* reports that the neural prosthetics enables leg-paralytic monkeys to walk again. The nano-swarm robot developed by the United States can easily deliver painkillers to specific body parts. In the field of genetic medicine, deep learning has offered possibilities for rewriting genes and synthesizing new organisms.

We don't want to discuss similar examples. Although nanorobotics and genetic modification are hot topics in film and TV, they are removed from the real world. More important, although artificial intelligence has entered many medical frontiers, it is still in the theoretical or experimental stage. But artificial intelligence has already played an important role in some more mature areas.

In the treatment of eye diseases, limited patient information causes 10 percent of patients to be injured during treatment, at least half of which could have been prevented. The famous Moorfields Eye Hospital in the UK has partnered with Google to create a machine-learning system that identifies eye-disease risks based solely on eye digital scans. Eye-scanning technology has long existed, but traditional machines cannot quickly analyze the complex eye data after scanning. Machine learning is expert in processing this type of data, greatly reducing the analysis time and improving accuracy, thus providing treatment solutions to patients in a faster and more accurate way.

IBM's Watson robot can help humans analyze difficult diseases such as tumors through big data and artificial intelligence. In Japan, most of the pathological readings required two doctors to prevent errors, but NEC has been implementing automatic reading systems in Japanese hospitals for many years. Now it has been able to replace the role of doctors to assist in reading.

Biological big data is also transforming the medical-research and pharmaceutical industries. In some traditional medical diagnoses, the doctor lets the patient walk for a few minutes on a smooth surface to record their walking distance. This test is to predict the survival rate of lung transplants, to detect the progression of muscle atrophy, and even to assess the health of cardiovascular patients. But if doctors want to get monitoring data for one thousand people at the same time, then recording will be difficult by not indicating other procedures.

Smartphones have solved data acquisition. A researcher in the United States wanted to collect data on cardiovascular disease, so he uploaded an app called My Heart Counter to Apple Store. In just two weeks, he obtained more than six thousand test results. Now what he lacks is no longer a sample but the ability to conduct more accurate analysis of large data.

Once we extract biological big data and arrive at conclusions, the entire medical industry will benefit a lot. In the case of diabetes, it is estimated that there are more than one hundred million patients in China, not to mention those in other countries. A problem that plagues the medical researchers is the accurate diagnosis of the subtype of diabetes that is extremely unfavorable for pre- and posttreatment. At this point, big-data analysis can play its advantage. If we can finally confirm the specific disease of each patient, we can truly achieve application of medicine according to indications.

Gene sequencing is a frontline method for finding the cause of disease. In this regard, I sponsored an esophageal cancer project. The research of gene sequencing and disease had been confined to single-gene pathogens due to limited technology; such research led to the discovery that a certain gene mutation can cause rare diseases such as Down syndrome. Many common diseases are caused by a combination of mutations in multiple genes. In the past, the computing power was not developed enough, so people couldn't figure out which combination of genes would cause common diseases at all. But this will be calculated in the future.

Big data and smart analysis will also significantly improve the pharmaceutical market. The variety of drugs in modern medicine has become increasingly complex, and sometimes patients find that there are hundreds of drugs to treat colds. Big data will bring us a customized drug-use plan to help us use the most appropriate drug.

This is the concept of precision medicine. Former US President Barack Obama proposed the Precision Medicine Initiative in his State of the Union Address in 2015. He proposed to analyze the genetic information of more than one million American volunteers to grasp the mechanism of disease formation, develop corresponding drugs, and achieve "precise drug application." With the help of the Internet, artificial intelligence, and biological big data, precision medicine will become a new way to prevent and treat diseases that takes into account differences in personal genes, environment, and living habits. In 2015, the Ministry of Science and Technology of the People's Republic of China held an expert meeting on precision-medicine strategy and plans to invest about 60 billion yuan (almost $8.5 billion) in precision medicine by 2030.

Professor Brendan Frey of the University of Toronto said that under modern technology we have been continuously accessing data on genetic biology, but it is difficult for humans to crack and control these massive amounts of data and to "understand" the genes. We should use deep learning to look for genetic connections that humans cannot find. When artificial intelligence finds healthy gene-sequence patterns, humans can genetically diagnose and even predict disease and optimize drug targeting.

The concept of precision medicine has caused a lot of controversy. Through genetic testing, American actress Angelina Jolie was informed that she had an 80 percent and 50 percent chance of developing breast and ovarian cancer, so she resolutely removed both breasts and her ovaries. Her action triggered a big discussion in the media. Many people would hesitate to have a removal operation just based on digital probability. More people are worried that predictive technology will be abused and become a means of "fake illness" to make money.

Predictive treatment is just like predictive prevention that AI has promoted in the field of police safety—taking precautions before a crime occurs. In any case, future humans must face such choices more frequently.

Besides leading to the sensational and controversial preemptive surgery, artificial intelligence is of great value in more common medical applications.

In China's big cities, more than 70 percent of the people are in suboptimal health, but due to poor access and a busy work schedule, only 5 percent actually go to the hospital. People need their own private doctor. The future medical brain will become this "private doctor." Everyone can "see a doctor" on mobile phones and smart systems at anytime and anywhere. People who are far away from big cities and grade-A hospitals will have more medical options.

There is no doubt that the future of the medical business will be changed through artificial intelligence. Fan Wei, the former director of Baidu's Big Data Lab, showed this—artificial intelligence has already allowed him to experience "time travel."

About thirty years ago, when he was a high school student, Fan hoped to become a doctor. He made plans and personally visited the admission office of Peking Union Medical College. However, for various reasons, Fan did not study there but went to Tsinghua to learn computer programming.

Interestingly, more than thirty years later, Fan really started research on medical issues. His current dream is to make a medical brain accessible for more people as soon as possible. A medical brain can automatically generate medical records for doctors, saving them time to treat more patients. With the help of the medical brain, doctors will definitely update their skill, and the doctor-patient relationship will be more harmonious. Fan hopes to use artificial-intelligence technology to turn the knowledge and experience of senior doctors into one piece of software for the benefit of more people.

11

SERIOUS NEW PROBLEMS IN THE BRAVE NEW WORLD

In 1932, Aldous Huxley, grandson of biologist and champion of evolutionary naturalism Thomas Henry Huxley, published the dystopian masterpiece *Brave New World*, depicting the future of the world of mechanical civilization. In the novel, people lived and worked in peace and satisfaction, but everything was standardized. Human beings were born in the incubation center, and they belonged to five different castes with different social status. Managers used "scientific" means such as test-tube cultivation, conditioned reflex, hypnosis, and dream therapy to strictly control people's personality and let them to perform their social roles and consumption patterns with a happy mood. Such a theme later re-emerged in works such as the movie *The Matrix* and became an inevitable topic for modern thinkers.

Neil Postman, author of *Amusing Ourselves to Death*, said, "In *1984*, people were controlled for fear of pain; in *Brave New World*, people were controlled by the blind pursuit of happiness." What will the future of artificial-intelligence society be like? Will it transcend the imagination of all people, form a new world beyond pain or happiness, or continue the theme of eternal struggle of mankind?

The intelligent revolution is the fourth technological revolution and comparable to the industrial revolution. Unlike the backward and passive old China in the industrial-revolution era, today, China is taking the initiative to control this revolution. Like the use of unmanned vehicles in challenging times, the era of intelligence becomes clear in chaos. We have reason to believe that China will occupy a commanding position in this transformation.

Thomas Henry Huxley (1825–1895), a British naturalist and educator, was a most outstanding supporter of Darwin's theory of evolution. His lecture "Evolution and Ethics," later published as a book, has profoundly influenced China's development because of Yan Fu's translation, which is called "Evolution of Heaven" in the Chinese version.

If we raise our sights higher and face the relationship between technology and human beings, we have some doubts. Careful readers may discern a concern underneath the enthusiasm of each chapter. We advocate artificial-intelligence education at the beginning because we are worried about the inequality at the educational starting point. We describe the new human beings in the future because we want everyone to master the survival skills of the intelligent age. We carefully imagine what the chief artificial-intelligence officer could do because we hope many companies will make collaborated progress in the era of great change and maintain economic and ecological balance. We know that artificial intelligence and the Internet of things will help China upgrade manufacturing, but we should also think about the new industrial crisis.

Throughout history, technological progress has always been in the hands of a few people and even become a weapon of injustice, and its popularity is full of twists and turns and uncertainty. Of course, the uneven march of progress partly results from selfish owners, but more is due to the slowness of the latecomers and the natural competition of the human race. Will AI play out the same way?

The development of mobile Internet in China is an interesting answer. Rural people use the mobile Internet much more frequently than PC Internet because the former is more available by smartphone. China's strong manufacturing industry has promoted the mobile Internet with the Internet infrastructure. So rural people are more equal. As a mobile resource, artificial intelligence and big data like the mobile Internet are also inherently penetrating and versatile.

Everyone will encounter AI with a different level of preparation. What kind of new world will AI create as a new tool? Much needs to be done. Chinese people once evoked a mighty hardworking revolution in silence, benefiting hundreds of millions of nationals and even the world. Are the diligent human beings positively ready for the intelligent revolution again?

Digital Gap

One of Wu Jun's sayings is very popular on the Internet: "The intelligent age belongs to only about 2 percent of all people, and the other 98 percent will become laggards." This figure appeared to us during the Occupy Wall Street movement that started in 2010. The slogan of the marcher was "against 1 percent." In the future, it is possible that people who hold financial capital are the 1 percent, and those who possess AI resources are the other 1 percent. This book will not discuss the movement but an amazing scene that happened shortly after it.

At the end of October 2012, Hurricane Sandy swept across the East Coast of the United States, where seawater flooded, and more than six million people lost electricity. The bustling New York City became an ocean, and 250,000 users in Lower Manhattan suffered a black out. But there was one single building brightly lit and independent—the Goldman Sachs building located at 200 West Street, Manhattan, New York City.

This scene caused public outrage. At that time, the Occupy Wall Street movement had been forced to end about a year earlier. The United States was still caught in the economic downturn caused by the subprime-mortgage crisis. The employment situation was sluggish, and people were full of complaints about Wall Street. Goldman Sachs is the chief representative of Wall Street's financial capital. The dark clouds fell down on the city, threatening to overwhelm it, and the shiny Goldman Sachs building stood in stark contrast to the surrounding black. A good symbol of this scene seems to be that the 1 percent was right there, even though the flood was all around.

But the Goldman spokesperson immediately explained, "We are not the only building with electricity, but we do have generators, and now we are completely powering ourselves." What he wanted to say is that the Goldman Sachs building was shining on its own strength.

This may also be a metaphor for modern society: some people stand at the top of the pyramid for various reasons; some rely on power, and others rely on capital, talent, or technology. Goldman Sachs itself is a financial company that is very skilled in data and machine intelligence, representing the combination of Wall Street and the high-tech industries in both the West and East Coast. Compared to the old infrastructure in the United States, Goldman Sachs has built immunity with rich resources.

When President Obama took office, the US high-tech industry developed at a high speed. However, by the time of presidential transition, the Democratic Party, which represents high-tech forces, encountered a Waterloo and lost the presidency and the two houses of Congress. Some senior commentators pointed out that in this election, the so-called white-protest phenomenon emerged around Trump. Whites who protested against the Democratic Party were white workers from industries other than colleges and high-tech industries. Some experts pointed out that the unrestricted expansion of US financial capital led to a serious imbalance between the virtual economy (related to the Internet) and real economy (related to manufacturing). The manufacturing industry had been squeezed out, and the combination of financial capital and information technology had exaggerated this imbalance, finally leading to a bad end.

Paul Krugman, a Nobel winner in economics, said the following with lament:

> In 2010, when a new railway tunnel was about to be built under the Hudson River, the governor of New Jersey suddenly cancelled the largest infrastructure project in the United States at the time. But meanwhile, another costly tunnel project was coming to an end: the Spread Networks tunnel which runs through the Allegheny Mountains of Pennsylvania. However, the original intention of Spread Networks tunnel was not to carry passengers, or even to transport goods, but to set up a fiber, which can reduce the communication time between the Chicago futures market and the New York Stock Exchange by 3 milliseconds. The railway tunnel project was cancelled, and the Spread Networks tunnel was built. Who cares about these 3 milliseconds? The answer is the high-frequency traders in the stock market who buy or sell stocks at a rate of a few milliseconds faster than their peers to exploit the difference in pricing. This phenomenon tells us what is wrong with the United States today. The society is allocating too many resources to financial speculation, resulting in a serious imbalance in the industry.

High-frequency trading, or flash trading, is a financial speculation method based on information technology, which requires comprehensive support of hardware, algorithms, and talent. Not only is the strategy of senior financial transaction experts required to be put into an algorithm, but also the algorithm has to be as efficient as possible, to obtain results faster than other algorithms. Furthermore, the algorithm should adapt to adjust the environment and adjust parameters and probabilities according to external conditions. This is already relatively developed machine intelligence. The independent Goldman Sachs is a master of this field, and it recruited a top computer team for this purpose. Similar high-frequency trading technology helps financial crocodiles leave the retail investors far behind, but it caused the stock market to collapse because of frequent machine errors and greatly increased the market uncertainty.

The lessons of the United States deserve deeper thought by the world. Digital technology has developed in the United States for a long time, and artificial-intelligence technology is the latest extension of it. Data and smart technology are not only the chief culprits to imbalances; the combination of unfair social systems and unscientific economic policies accelerate imbalances. The ordinary people are the main bearers of the consequences of imbalance.

The sharp contrast between the eastern and western states on one hand and the middle states and Rust Belt in the northeastern part of the United States on the other hand in the 2016 election portrays not only political opposition but also the digital divide, to some extent.

Of course, the gap is not entirely caused by digital technology; it just deepens it. In 2014, French economist and professor of the Paris School of Economics Thomas Piketty published the book *Capital in the Twenty-First Century*, which was very popular for a time and highly recommended by economists such as Krugman. The book proves with convincing data that in the past few decades, income inequality in the world has expanded, and the return rate on capital has greatly exceeded the return on labor of ordinary people. The unequal production relationship described by Karl Marx has continued, without any signs of improvement. The United Nation's report "China National Human Development Report 2016" shows that for the past decades, the major global poverty-alleviation achievements have occurred in China, with other countries being lackluster.

Marx believes that the purpose of capitalist industrialization is to maintain high profits, rather than directly promote the well-being of humans, thus resulting in a large surplus population or unemployment. Workers are waiting to be hired to replace workers who have lost the value of exploitation. But in the age of intelligence, the meaning of the surplus population may change.

What Else Can Humans Do?

Many people have not realized that all single-skilled occupations may be replaced by machines. Speech recognition will replace stenographers in the next few years; simultaneous interpretation will take some time because of the limited development of microphone array hardware. Electronic police are more effective than security guards. Camera recognition has already replaced card-swiping in many residential districts. The machine has replaced manual customer service of the e-commerce and express delivery industries. However, taxi drivers may disappear later, and unmanned vehicles may find it hard to replace human drivers within the next ten years.

Amazon's unmanned supermarket has excited shopaholics. Here, as long as the customer installs a specific app in his or her phone, the customer doesn't have to wait in line or swipe a card at the counter; instead, the customer just carries out the goods. The sensor automatically identifies the item the customer took, calculates the price, and automatically debits the bank account when the customer leaves the supermarket. Some people joked that the shopping experience is like free robbery, which satisfies their desire.

This is of course an upgrade to the consumer experience. Paradoxically, smart supermarkets, while benefiting shoppers, also make cashiers lose their jobs, which in turn reduces their shopping ability. It seems like the wave-function collapse in quantum physics—a superposition of states is found in the form of a wave, but during the act of observation, the quantum system must exist at a certain position, so the original quantum wave collapses to a point. The quantum system exhibits practicality, so people can observe it. The relationship between technology and happiness is the same. When you see happiness, something is failing behind.

In addition to cashiers, white-collar workers such as lawyers and news editors who happily shop here may not be lucky either. Their positions may also be replaced; in some companies, machines have performed at least basic legal analysis and draft writing.

At the end of 2016, Russia's largest bank, Sberbank (Russian Federal Savings Bank), announced the launching of a robotic lawyer to handle various complaints. This resulted in about three thousand legal professionals working in the bank being fired. Vadim Kulik, vice chairman of Sberbank at the time, said that all future general legal-document processing would be automated; lawyers would only be needed for the task in urgent legal proceedings.

In the future, maybe only some nonstandard work that requires intuition and creativity, such as art design, planning, and organizing, will be irreplaceable.

Losing the purchasing power might not matter too much, because the shopping itself may disappear too. Marx believes that exploitation leads to poverty in the working class, which in turn causes low consumption and insufficient demand, which eventually results in overproduction and collapse. In the intelligent age, efficient machine production can meet the basic needs of more people. Factory owners may not even be willing to exploit people because robots can do better than human workers. People who lose their purchasing power can also get basic benefits from the abundance of machine production.

New jobs are indeed created for humans, such as unmanned vehicle traffic management, or Amazon's fifty thousand data-tagging workers.

Ordinary people won't have to worry about food and clothing. Can they realize the free life, as Marx put it, "to hunt in the morning, fish in the afternoon, rear cattle in the evening, criticize after dinner"? Maybe, but they may lack freedom because they can't create value, and they don't need it. One of the basic needs of people is recognition. How depressed will a valueless person feel? Perhaps they will regroup in a dungeon and have a different lifestyle than that of the smart world.

The dungeon is a bit far away. In the short term, the increase in the surplus population will be a real problem that may cause social unrest. We can imagine that there will be new law provisions for labor protection in the future, such as stipulating that the ratio of robots and labor employed by each enterprise should reach certain standards. But the priority may be to upgrade social vocational training and provide vocational education in computers, the Internet, and artificial intelligence for ordinary workers.

The US government has also recognized this challenge. In December 2016, the White House released the report "Artificial Intelligence, Automation, and the Economy," which provided strategic recommendations to respond to job market changes, such as the following:

- Invest to educate and train Americans for jobs of the future.
- Aid workers in transition, and empower workers to ensure broadly shared growth.

This proposal sounds good, but it remains to be seen whether the US government can effectively cope with the intelligent revolution, considering that it is still struggling even in the implementation of health-care reform. John Hopcroft, a professor of computer science at Cornell University and a Turing Award winner, predicted in early 2017 that with the rise of artificial intelligence and automation technology, the size of the US workforce may be reduced by 50 percent in the future.

What can China do to amend inequality? In the digital age, can we use technology to narrow the gaps? We are just here to inspire others, hoping to attract more far-sighted people to think and explore. The brain of AI education should be put into vocational training, so that machine-assisted ordinary workers learn to deal with AI and dig out more value in the AI era.

The government should introduce corresponding safeguards, but the key is to improve the production methods themselves. The US manufacturing industry is hollow; high-tech industries fail to feed the manufacturing industry but combine with Wall Street capital to find profits, which has deepened the digital divide in the United States. China can create a new balanced environment with an overall initiative on remolding its developed manufacturing industry by correct guidance and support.

The famous Hungarian political economist Karl Polanyi pointed out in the book *The Great Transformation* that capitalism and industrialization digitize and rationalize everything and squeeze humanity, especially to put the market above the society rather than letting the market be embedded in it. The market drives away all social relationships. For example, the enclosure movement in the United Kingdom, referenced by the "sheep devour men" proverb, essentially expelled the farmers and turned the farmland into a grazing area to produce wool commodities, thus destroying the rural social order. All of this caused a humanitarian disaster and led to a reverse social-protection movement, including the fight for welfare and the pursuit of labor-protection laws. After centuries of turmoil and adjustment, the Western capitalist countries reached the basic balance of production and welfare.

The historical experience and lessons are worth learning. Today, we find this balance is once again in crisis. More and more people are turning their eyes to China. Both strategic scholars like Martin Jacques and technical experts like Kevin Ashton are positive about China's development. China's efficient government and the people's strong tradition of pursuing development and fairness offer new possibilities for this.

Famous economist Justin Lin Yifu insists that an "effective government" is the guarantee of social development. In the face of the intelligent revolution, the "efficient government" needs to be prioritized, focusing on the competition and upgrading in key areas. However, the areas related to the national economy and the people's livelihood need to be balanced to minimize social fluctuations during the transition period. During the reform of state-owned enterprises in the 1990s, the Chinese government, in conjunction with social forces, had established a relatively mature reemployment training system. Today we need to upgrade this system to guide ordinary workers to be familiar with smart technology and to apply what they have learned. Give full play to the role of community offices, and open smart-life lectures. Give students the opportunity to learn smart technology even at the basic compulsory education stage.

Large companies must also consider social responsibility in the development of technology. Technicians need to think more about social responsibility and try to embed AI technology into social development, instead of just putting the development of AI itself above society. There have been good attempts in this area, such as helping people in backward areas to read, through intelligent technologies on speech and vision; helping people with disabilities; and designing products for labor training. In these respects, technical workers need to think more; for instance, *Can an online-education platform for college students and technicians provide training for the general labor force?* All of this requires the cooperation of the government and the enterprise.

A Question of Instrumental Rationality

With technological advancement, people have no choice but to move forward. Technology is not an outsider; it comes from human production and creativity itself, which is inseparable from human existence.

Philosophers have discussed the relationship between technology or tools and humans in detail. Marx first acknowledged that capitalism and industrialization have created more than the total amount of wealth created by mankind in the past few thousand years and have generally improved human life. But then he talked about "alienation," meaning that what humans created in turn enslaved some humans. Another great thinker, Max Weber, argued that the monetization of capitalism and its calculation, when dealing with all things and relationships creates the so-called instrumental rationality, which was originally intended to give humans a "light cloak," but ended up being an "iron cage."

Instruments liberated mankind from natural tyranny, but new inequalities emerged from the combination of the instruments themselves and tools with social implications. Whoever monopolizes these tools monopolizes the life-blood of economic politics. The society split into two classes due to the application of tools: the ruling class, which monopolizes intelligence, and the ruled class without any access to information.

The steam engine accelerated the industrial revolution. As early as the Ptolemaic Dynasty in Egypt, temples used steam power, but only to enhance the statues with mysterious movement. This is the epitome of the monopoly of intelligence. Only after the combination of capital and intelligence could the steam engine fully realize its potential and inject a strong motivation into civilization.

The industrial era separated intelligence and labor still further. One of the major drawbacks of mass manufacturing is the separation of thought and behavior. Innovative tasks are carried out by scientists and engineers far away from the production line, while production is done by unskilled workers.

Cloud society and the Internet of things offer the possibility of combining practical and mental work.

In the traditional industrial system, workers are "standardized" as tools, especially workers on the assembly line, who can only passively follow the instructions to complete the prescribed actions. Here "work" is alienating: it is no longer a creative an action but a negative mechanized action. We can boldly imagine that future artificial intelligence may provide technical solutions. For example, considering the diversity of the labor force, we may create an intelligent production process that transcends the assembly line by automatically adapting and adjusting according to the labor situation without affecting the overall situation of nonstandardized workers. It could provide the workers individual opportunity. The company's intelligent operating system would connect frontline workers and researchers with managers through the network. Worker's operating habits and status are recorded, located, and analyzed at all times, and optimization suggestions are given to synchronize work and study.

China's complete industrial chain, diverse work scenes, and rich talent are all valuable assets. Scientists and engineers should consciously go deep into production scenarios and combine smart technology with industrial practices and life practices.

Useless Use of Robots

Artificial intelligence conveys to us an inspiration beyond instrumental rationality.

Both Astro Boy and Baymax, robots from the childhood of many readers, surpassed the tool attributes originally set by humans, and they strove to pursue their own independent "humanity."

In *The Iron Giant*, in order to save the human village, the robot chose to die together with the atomic bomb. At the last minute, it said, "I know what I choose to be: Super Man!" Whether the robot itself can have humanity or even surpass humanity is an interesting and meaningful topic.

Domenico Parisi's *Future Robots: Towards a Robotic Science of Human Beings* is about not just robots but also a way to reflect on the instrumentality of human science and point out the direction of human beings. A fundamental feature of science is objectification, or the separation of people from their own research objects, enabling the observation, recording, and analysis of objects without observer's own emotions, positions, and so on getting in the way. This method is a powerful tool for human beings to understand the natural world, but it makes understanding humans quite difficult.

First of all, researchers cannot completely separate themselves from the research object. People have their own emotions, motivations, desires, and cognitions.

Second, the research theories about human behavior and human society are basically expressed in words rather than the mathematical symbols used in natural sciences. As a tool for expressing scientific theories, words have great limitations: the meanings are inaccurate and they reflect the researchers' subconscious values and emotional orientations, which are difficult to digitize. Words are not a transparent intermediate tool, and the words themselves construct a layer of "reality." Words are even numbers, and their essence is symbolic. There is always a gap between symbols and reality, just as the limited vocabulary about colors can't fully express the infinite variety of detailed colors in reality. Words may even lack realistic counterparts, such as *belief* and *target*.

The development of big data and mathematical methods has enabled us to digitize many anthropological phenomena, but it is still not enough to study humans. Parisi said, "[These methods] have not even tried to identify the mechanisms and processes behind these behavior and social phenomena, let alone to explain them."

Artificial intelligence offers new possibilities to express scientific theories about humans. Parisi proposed "theories as artifacts"—creating artificial structures (humanoid robots) to replace research. This theory is based on the principle that "no matter what X is, to understand X, it must be reproduced manually." If your theory is not good, then it cannot be built by artificial construction. If you can make a humanoid robot, that means your research is successful. Fundamentally speaking, this is a unification of research and practice: research is practice, and practice is research. In addition to overcoming the aforementioned human issues, this process can also integrate various disciplines such as sociology, psychology, biology, political science, linguistics, economics, literature, history, and philosophy to break the current situation of mechanical segmentation among various disciplines.

Parisi believes that the current robot or artificial intelligence is different from human beings in that human thinking includes both motive and cognitive modes. Any act of a person has motives, such as eating and drinking for living, reproducing for safety, ideals for honor, etc. The existing robots only have cognition ability but no motivation because people set its purposes. Robots are just tools to meet the specific needs of people.

But robots can be a shadow of a human, for which Parisi categorized two types of robots: practical robots and humanoid robots. The latter is exactly what scientists need. The former has only cognitive thinking, while the latter is set with motivation. For example, a robot is set to acquire energy regularly to survive through a robot program in the computer, which can be equipped with a starvation sensor to look for food and water in a certain virtual space. The food is replaced by a food token, the water is replaced by a water token, and the two tokens are placed in different areas. Then the robot decides the route and order of obtaining the tokens autonomously and will die if it fails to find them. We observe how the robot makes decisions.

Similarly, robots can be set to circumvent some dangerous tokens to ensure safety. They need to exchange new codes with other robots to create new generations (reproduction), and even have property concepts and motivations to follow rules. Humans observe how robots would evolve to have social humanities, language, culture, emotions, and even art and religion under the drive of these motives. For example, male robots and female robots are endowed with the motivation to protect offspring robots, and then we observe how they develop familial concepts.

Such types of robots made by scientists may not have any practical application value—those robots could make mistakes, get sick, dream, and be sad or happy. Instruments should not make mistakes, and these are all human characteristics. Under this way of thinking, artificial intelligence would no longer be a tool to serve the external needs of human beings but instead a means by which humans understand themselves. Such research has already yielded some results, opening up a vision that has never been seen in psychology and sociology in the past.

However, readers may sense danger; will the humanoid robot eventually evolve self-awareness? How would it deal with human beings? Would religious people regard this research as an arrogation because scientists would be doing what God did—reproducing people? Scientific research is different from religion, and its essence is the objective study of human beings. Life is constantly generating and creating. Science would encounter a new scene here: the realization of instrumental rationality in robots. In the final analysis, the research would carry out self-reflection through artificial intelligence while at the same time guide humans in the development of artificial intelligence, which inspires us to understand how robots could be integrated into human society and how they could participate in human culture and social life.

Protestantism sees human beings as a tool of God to achieve divine will, and human self-reflection seems to always overcome this will. People should not be just tools but instead be the unity of goal and means, which can be put into practice with artificial intelligence.

Catch-23

It seems necessary to set ethics and rules for humanoid robots.

At the 2017 Asilomar Beneficial AI conference held January 5–8, 2017, 1,980 AI-related experts from industry, academia, and research circles jointly formulated the "Asilomar AI Principles," which aims to ensure a healthy development of artificial intelligence in which everyone can benefit. These twenty-three principles are divided into three parts: research issues, ethics and values, and longer-term issues.[18]

In terms of research issues, the goal of artificial-intelligence research is to create beneficial intelligence, rather than undirected intelligence, as the biological evolution. The investment in AI should be accompanied by funding for research on ensuring its beneficial use, including thorny questions in computer science, economics, law, ethics, and social studies, such as how to maintain a high degree of robustness[19] for future AI systems, so that they can do what we want without malfunctioning or getting hacked. How can we grow our prosperity through automation while maintaining people's resources and purpose? How can we update our legal systems to be more fair and efficient, to keep pace with AI and to manage the risks associated with AI? What set of values should AI be aligned with, and what legal and ethical status should it have?

Other research-related issues include constructive and healthy communication between AI researchers and policy makers; fostering a culture of cooperation, trust, and transparency; and avoiding any race and cutting corners on safety standards.

In terms of ethics and values, the twenty-three principles provide lots of visions, including the following:

- AI systems should be safe and have failure transparency. If an AI system causes harm, it should be possible to determine why.
- Any involvement by an autonomous system in judicial decision-making should provide a satisfactory explanation auditable by a competent human authority.
- Designers and users should bear the responsibility of moral influence.

---

18 Translation by the AI Era, an artificial-intelligence industry think tank.
19 Robustness of a robot is a property that describes the ability of a robot to maintain its original or expected state when it is disturbed or even changed internally or externally.

- Highly autonomous AI systems should be designed so that their goals and behaviors can be assured to align with human values throughout their operation.
- AI systems should be designed and operated so as to be compatible with ideals of human dignity, rights, freedoms, and cultural diversity.
- People should have the right to access, manage, and control the data they generate, given AI systems' power to analyze and utilize that data.
- The application of AI to personal data must not unreasonably curtail people's real or perceived liberty.
- The economic prosperity created by AI should be shared broadly, to benefit all humanity.
- Humans should choose how and whether to delegate decisions to AI systems, to accomplish human-chosen objectives.
- The power to control the highly advanced AI systems should be given respect and improve, rather than subvert, the social and civic processes on which the health of society depends.
- An AI arms race in fatal autonomous weapons should be avoided.

About long-term issues, the following principles apply:

- It is suggested that with no current consensus, we should avoid strong assumptions regarding upper limits of the future AI capabilities.
- AI should be planned for and managed with adequate care and resources to avoid catastrophic or existential risks.
- AI systems designed to recursively self-improve or self-replicate in a manner that could lead to rapidly increasing quality or quantity must be subject to strict safety and control measures.
- Super intelligence should only be developed in the service of widely shared ethical ideals, and for the benefit of all humanity rather than one state or organization.

The twenty-three principles remind us of *Catch-22*, an example of gallows-humor literature, where the twenty-second rule is impossible. These twenty-three principles are more like twenty-three appeals, which raise questions rather than solutions, and represent desires rather than realities that already exist. The principles focus a lot on values. It is natural that scientists advocate for the development of robots to follow human values. But here are three points to consider:

- The values here are based on the fact that machines are only tools for betterment, and they should not harm human interest when being used. In fact, humans cannot guarantee this in any tool, such as armaments. However, it is indeed possible to limit it as much as possible, such as with an arms treaty.
- People are more serious than ever before because they realize that these robots are different from previous tools. Ordinary tools have no life or consciousness and completely obey human operations. But if robots have a humanoid consciousness, how will humans treat them? Obviously the twenty-three principles did not consider this issue. Humans refuse to enslave human beings; it is a slavery to use humans as a tool. Then what about robots? If a human value is to make robots a tool, will humanoid robots consider that value dignified?
- There are various conflicts in human values. As we all know, the participants in the formulation of these principles are basically from the Silicon Valley. The attitude and values of the Silicon Valley camp in the 2016 general election have clashed with the Trump camp. The rules they set seem to be beautiful but may not meet the values of many people in reality.

The twenty-three principles express the good wishes of mankind, which is equivalent to a set of political correctness formulated by scientists for robots. Considering the embarrassing situation of political correctness in the current United States, we should understand that the political correctness of artificial intelligence will not be achieved overnight. Signing off on the principles is one thing, and implementing them is another. Values are not static ideas but are generated in the game. Values cannot exist in isolation in a certain group. For example, the values of robots are actually the values of humans. In the future, AI may join this game to form new and dynamic human-machine values. We need to constantly explore the healthy road of AI and human development in practice. The twenty-three principles serve as a pathfinder.

Realistic Legal Issues

Although the prospect of the algorithm replacing the law has cropped up for many scientific and technological people, the law is still the main means of regulating human social relations. We need to consider the challenges the rule of law will encounter in the era of artificial intelligence and how to deal with them.

Artificial intelligence is not just about a technology but also about changing the way the society operates. Movies such as *Minority Report* have foreseen preventive management. The biggest role of deep-learning technology is prediction, which also brings new inspiration to traditional legal thinking. For example, the law will transform from the compensation mode to the prevention mode. However, the process involved is very uncertain.

In terms of the complexity of the algorithm, the technical community defines strong AI and weak AI (or narrow AI). Some scholars believe that artificial intelligence can be defined according to technical cognition, but it is difficult to legally develop precise standards according to the degree of intelligence. The way the law deals with the complex world is to establish general simple rules, conceptually abstracting social norms (such as behavior) rather than targeting specific circumstances; otherwise, the rules would become extremely complex and difficult to understand and operate under. The information among production materials becomes more and more transparent, but as the information-matching intermediary, AI becomes more opaque and its rule design and operation more difficult to be understood by users and even developers. This all goes back to how the law deals with code.[20]

The problem exists in the conflict between the legal abstraction and the black box of technology. For example, in the Qvod case (concerning pirated videos) involving Internet technology rather than artificial intelligence, the process of liability determination was long and difficult. Regulators or courts do not want to go deep into the algorithm to understand the technical cause of an accident. If the law determines that the black box should be controlled within reasonable limits and the accident could have been avoided, then the black-box provider should be held responsible. We foresee that insurance (or

---

20 Hu Ling, "Legal Imagination on Artificial Intelligence," *Beijing Cultural Review*, no. 2 (2017), 108-116.

even compulsory insurance) is a good choice for markets with low probability of accidents but huge potential losses. Technology-involved markets of aviation and medical insurance are well developed, and insurance will likely extend to more service industries driven by artificial intelligence.

On the other hand, it may be necessary to rely on the top-level design of the algorithm to prevent negative consequences. Artificial-intelligence technology will not be a field exclusive to science and engineering professionals; legal professionals and other administrators also will need to learn about AI, which will provide the technical requirements to legal practitioners and other administrators. Rule-of-law management needs to be integrated in the production process, such as the management of data and how resources are processed by the algorithm. For example, the credit-information system will prohibit the collection of discriminatory information, such as race, gender, religious affiliation, political inclination, etc. In the word-of-mouth network, suppliers must fight against false data. Network taxis and drivers should face regional supervision. The legal robot itself is also a good source to help people make judgments against complex rules—to solve technological problems through technological advancement.

There may be plenty of AI-related issues. What follows is a brief discussion about privacy issues that we care about in everyday life.

Some people say that in the era of intelligence, people live in a society without privacy, because all data is in communication, and even the heart rate is shared through wearable devices. The future paparazzi will also be intelligent, finding traces of the stars through data analysis and ubiquitous visual surveillance. How should we protect privacy in the smart society? We may have to break through the traditional concept. Privacy is of course part of public consciousness. In reality, people emphasize privacy protection but at the same time they are keen on recording and even spreading their private information. There is no individual without privacy, but in the data age, absolute privacy prevents individuals from communicating and being identified. The best way is to update the system and technology, such as establishing a unified data-protection platform so that individuals can understand the status of their data being used by the government and company to avoid excessive use.

Redistribution of Digital Power

The late famous Polish thinker and sociologist Zygmunt Bauman believed that the digital age has brought about a problem thornier than monitoring: separation of power and politics under technology. Traditional politics tries to solve problems within a country. However, transnational capital and its political power create more uncertainty in various fields through technology. At the same time, ordinary people's trust in technology and bureaucracy has been erased through this conflict. Power flows with capital and digital networks, and traditional politics is powerless to bind them. With the rebound of global capital power, the tribalism of European and American countries has reemerged. This is the background of the rise of Brexit and isolationism in the United States. Bauman described Western countries, but uncertainty is also common in the new era.

Political, economic, cultural, and media powers are infiltrated by data intelligence. It is not difficult to understand that financial capital has used the digital network, smart investment, flash trading, and other technologies to influence the world and make a big sensation. Social media such as Twitter have catalyzed the turmoil in the Middle East, and digital power has crossed the boundaries of sovereign states. Even the United States, which used to export digital power in the past, has begun to worry about being attacked by a multinational digital power. Julian Assange and his WikiLeaks, as a non-governmental organization, repeatedly exposed negative information about governments, which is an embarrassment. Hackers have gained the ability to deter the American ruling class with network technology. The US government even accused Russia of using hackers to intervene in the 2016 elections. This tension is unprecedented. This shows that no one can be spared of digital power, but it also indicates that it is necessary for individuals and countries to adapt to the digital intelligence era.

The form of digital power is like the modern power described by Michel Foucault. It is not a centralized and circular structure but a complex and multi-centered network structure. Digital power is scattered in this network and is elusive.

At the national level, it is necessary to establish the top-level design to prevent digital power abuse by the government. In reality, digital powers have maliciously challenged the normal functioning of the state and society.

Bloomberg Businessweek reported in 2014 that a Colombian hacker claimed to have manipulated voting in nine Latin American countries by stealing data, installing malware, and forging large-scale public opinion through social media. His team installed malware at an opponent's camp Colombian headquarters to monitor computers. In this way they obtained a lot of documents, such as speeches, meeting plans, and election arrangements. Based on this information, he used the virtual army to publish fake tweets and use a low-end Twitter robot to increase popularity and fans. Programs were set to automatically dial through the campaign phone in the early hours of the morning to harass the voters in the name of the opponent's camp. In a similar way, he has influenced the democratic elections in Venezuela, Nicaragua, Panama, Honduras, El Salvador, Colombia, Costa Rica, and Guatemala to varying degrees for eight years. He said, "When I realized that people are more willing to believe in online public opinion than reality, I found that I have the ability to make people believe in anything."

On Chinese and foreign social media, the simple and intelligent machine Internet water army (originally, paid posters who leave positive or negative feedback for a product) is ready to make trouble. It has been said that in the future, you won't be able to tell if you are talking with a person or not on the Internet. This has already come true.

However, the water army is not necessarily negative, because it could be used for the correct purpose. In a sense, Q&A robots and machine customer service are variants of the water army. Even in the forums and comment areas where dialogue tends to be meaningless, robots may perform better than humans. Wang Haifeng said that in many comment areas, human beings often curse or free-associate, while robots express meaningful comments that heat up the comments and maintain a positive public opinion.

Every modern person is surrounded by data he or she produces. Intelligent data has become the second body of human beings. Just as we are worried about the first body's safety issues, such as illnesses and car accidents, the second body also faces security risks.

Personal information faces various risks of disclosure in the Internet age. The technology of the fraud industry is also constantly being updated. The criminals even use the big-data method instead of common phone fraud. Police in the Shanghai Huangpu District cracked a horrendous online theft in early 2017. The hackers first used software to generate phone numbers in bulk, then used scanning tools to upload these numbers one by one on the Internet. Through some websites where hackers stored the leaked passwords (generally called the "social worker library"), the login password corresponding to the phone number is found. This is known as to "hit the library" in the industry. With this crude method, hackers could quickly get login information about many users and then use it to conduct criminal activities, such as stealing bank deposits.

There are always criminals in a society. The explosive growth of data technology in the Internet era and the backwardness of human adaptability have brought loopholes for criminals. This is like the new traffic insecurity that was brought to society by the just-invented automobile, but then people invented traffic lights and adopted new rules to control technology to ensure safety. Traditional data-management methods are riddled with holes in the Internet age. It is difficult for common people or groups to actively protect their privacy data. This requires the joint efforts of many parties. The police can also use big-data technology to deal with crime, and technology companies can help.

Now we have traffic lights in the big-data field. Taking telecom-fraud as an example, many companies, including Baidu Security, have partnered with the government to combat telecommunications crime. A safe-number library and scam map are built on big data, to share the information with public security organizations to locate the number and location of a fraud suspect. Additionally, a fraudulent phone number can be intercepted in real time and displayed synchronously on the fraud map, identifying fraudulent phone dynamics and reminding users with red lights. Baidu Security has accumulated more than two hundred million security numbers and blocks harassment and fraudulent calls more than one million times a day. The scam-phone interception rate has remained above 99.98 percent. At the same time, users having security software installed on their mobile phones can update the reported numbers and mark suspicious calls to the server to help identify fraudulent numbers. In this dynamic loop, security software will become smarter, and the ability to identify fraudulent numbers gets stronger.

Deep learning has also upgraded its firewall technology. Webmasters know that every time they install protection software, they have to set a lot of rules to prevent unauthorized access. Deep-learning technology allows the security system to automatically learn user access rules and identify abnormal access. The system sets its own rules, which are far more efficient than human network management.

Although some malignant fraud cases and security incidents have been exposed and invoked fear, we must know that more digital fraud has been blocked, and it is AI technology that provides support—both the network financial-security technology and smart city's intelligent monitoring technology. When Satan covers one foot, God covers a mile: they are constantly progressing in the game.

More often, the technical geeks will use their ability to make some "joyful" cross-border behaviors. For example, when Alipay had just launched the AR Red Packet (a red packet, or red envelope, is a monetary gift given for holidays or special occasions in some Asian countries), some people trained the complex neural network, reversed the clue photos that were covered by the horizontal strips, obtained the original image, and easily got the red packet. The method was not difficult, but the user had to know how to use the deep-learning sharing platform.

Because of traffic and pollution, license plates in Beijing and Shanghai are particularly difficult to obtain. A bimonthly lottery awards the plates. The license plate winning number is determined by the system "randomly," but this randomness follows a certain algorithm. Some people use machine-learning technology, which is alleged to have increased the success rate.

Looking around your working environment, you will find that some people are good at using a lot of technical tools to strengthen their abilities, yet some are tired of remembering the various login passwords all day. In the future, the familiarity with data intelligence will affect the happiness of a person's work and life. Only prepared people can enjoy data life.

New Generation and New Future

Finally, let us turn our attention to young people, who are the most valuable asset of the future human society.

In many developed urban societies such as Japan, China, Europe, and the United States, a group called NEET (not in education, employment, or training) has emerged, mainly referring to young people who are unemployed, stay at home, and are addicted to animation. According to statistics, there are over six hundred thousand NEETs between the ages fifteen and thirty-four in Japan, accounting for 2.2 percent of the population of this age group.

NEETs are often addicted to the 2D culture, highly coincident with the so-called otaku category. The erotic and violent yearning for animation games and the fear of the real world are the most basic characteristics of this group. They can't resist the vivid and cute animated characters on the screen. When faced with an AI character or even a physical robot perfect in all aspects, how can they control themselves against such irresistible products?

In June 2016, Andrew Ng and Liu Cixin had a conversation about AI. "Big Liu" put forward his own bold idea for the future AI: AI can become a human sexual partner and fundamentally change the cultural trend of human beings.

Can AI completely replace the role of a person in life? Ng, being a scientist, was more cautious, saying that AI is still far from technically maturity, but he does not deny the possibility of such a future.

In the future, human beings will create all kinds of "perfect" things according to their own wishes. But would that be really perfect? Would it be just a mirror of human beings?

When artificial intelligence breaks out, humans may keep themselves in the virtual world. Liu Cixin wrote a short science fiction novel in which humans connected the brain to the virtual world and became obsessed with it. They felt a God-like illusion and no longer investigated the universe, thus the earth civilization was terminated.

Like all new technologies, artificial intelligence promises a better life. It makes life more comfortable for many people. But humans still need the spirit to struggle. Sun Di, former economist and former general manager of China Construction Bank, who first translated and introduced Buffett's investment philosophy to China, believes that it is wrong for Western economics to rationalize individuals as the basis for economic development. Those who dare to innovate boldly are people with animal spirits. The society based on artificial intelligence can make people's life more refined, but people may instead be bound by various sophisticated dogmas and lose the brave animal spirit. In the future, everyone will easily access the smart flow, but now, we must think about where to apply the smart technology to meet the people's value.

Whenever a new era comes, there are always some people gearing up, some drifting, and some feeling overwhelmed. In recent years, people often have said that "the future has come, but it is not popular yet" with optimism. But we must realize that there have never been completely beneficial changes; they must be accompanied by loss, different perspectives, and shock, only with slow adjustment to the new way of life. Although artificial-intelligence scientists are at the top of the technology pyramid, their attitudes toward action are generally conservative. Zhou Jie, who previously worked in Baidu's deep-learning lab, said that his attitude is more "conservative," which does not mean stagnating but avoiding an overly aggressive approach for the direction of AI development.

More than one hundred years ago, Yan Fu introduced the translation of *Evolution and Ethics* to China. In his translation, he deliberately distorted the original work, emphasizing only the competitive side of biological evolution, while ignoring Huxley's thinking about ethics. Being in England at that time, which was the first power in the world, Huxley dreamed of a more benevolent England for his people, and Yan Fu, who faced "the biggest change in the past three thousand years," emphasized the law of the jungle and used "survival of the fittest" to summon a resurgent China. Time has passed. Today, as the Chinese stand on the road toward ascendancy, they must inherit the sages and surpass them at the same time. Like Huxley, we should think about the power of a smart society as well as how to make a smart society more harmonious. Anxiety does not mean pessimism, nor is optimism only true when it is based on sorrow. Imagining the future is a difficult thing, and although the future is unparalleled, it is worth fighting for.

In the many smart years ahead, even artificial-intelligence companies such as Baidu and Google may be just a visitor in the vast realm of history. Human beings have their own weaknesses and strengths, both in shortsightedness and ambition. The ancients said, "Not pleased by external gains, not saddened by personal losses" and "Get things done here and now, not caring about the future comments." This is the embodiment of our human spirit. What we can do is seize the opportunity. The existence of mankind is "on the road." Baidu wants to lay the foundation for the beautiful new world. China is going to transform from a big country to a great intelligent and civilized country. Everyone should not be left behind by the machine. Try to be a better person—know more, do more, be more, and work together toward a better but uncertain future.

POSTSCRIPT

Sometimes it is harder to describe the present than to describe history—especially when on the verge of an upcoming revolution. Yes, the AI revolution—we are sure that this will be a great change. As a technological trend that originated from genius and transcendence, it has been in existence for more than sixty years, and it has accumulated much energy through its ups and downs.

The advent of the data world has made it possible to directly verify the different genius algorithms and formulas in the scientific community that had been abandoned, left out, and even only suspected. Artificial intelligence, the dreamlike "concept machine," suddenly received a steady stream of fuel and power to keep it running at high speed.

More important, the industry has begun to take over the torches of the academic world. The cutting-edge intelligent technology once full of sci-fi imagery has really begun to step out of the laboratory and into the lives of ordinary people.

Fortunately, as a search-engine company, Baidu, since its inception, has everything for the development process (big data→deep learning→extraction→user value) and development culture. Perhaps because of fate, Baidu is the global leader and one of the most determined practitioners in the field of artificial intelligence. We are witnessing the magical energy and broader prospects that artificial intelligence has shown in the search environment. If the main theme of human progress is to know more, do more, and be more through the continuous improvement of perception and cognition, then artificial intelligence is the latest repercussion of this main theme.

However, in the face of a very exciting and unpredictable future of intelligent society, we are also worried. Although we are convinced that artificial intelligence will bring earth-shaking changes and great value to the society and all walks of life, we still cannot accurately describe such changes at this stage, and we may not be able to present such values in a short period.

How can a labor-intensive manufacturing company realize the intelligent upgrade of its products? How can a large farm achieve truly refined agriculture? How can a financial company prevent risks and increase revenue? How should a pharmaceutical company keep up with the future of personalized medicine . . . ?

Many industries have resorted to Baidu after confusion over intelligent upgrades. Some of those problems can be or have been explored by Baidu, and many of them are beyond Baidu's ability. Because all walks of life have their own evolutionary laws and the understanding, the knowledge maps that urgently require AI depiction differ greatly.

So, we often feel powerless when we try to portray the intelligent transformation of all walks of life in one book. Only experts might be able to correct the inevitable omissions and mistakes.

But this book may have been needed, in that by opening up our own capabilities and describing them in as much space as possible we can attract more potential collaborators. Thus, we can lay a foundation for the perfect match of "you have a story, I have wine" future cooperation.

We believe that for traditional enterprises that need to be transformed and upgraded, the first task is to clarify the standards of artificial intelligence. Perhaps history has given the Chinese economy an opportunity to remove the low-end knock-off image of Chinese exports and truly lead the smart industry, rather than being complacent. With authoritative standards, traditional enterprises can gradually find the coordinates of their own intelligent transformation and prevent themselves from going astray at the beginning.

Thanks goes to Robin. You have determined that when the intelligent revolution is about to come, we should open Baidu's smart environment to users and partners, popularize basic knowledge, and clarify the standard of specifications, which is also the original intention of this book. Besides, I'm happy and surprised to have you write some important chapters in person after you had set the logical structure and basic context of this book. As Baidu's chief artificial intelligence officer, your description of the future smart age is fascinating.

POSTSCRIPT

Although you come from different systems, you became the artificial-intelligence "problem raiser" for the intelligent upgrade of various businesses. You have improved Baidu Brain's ability to "listen," "see," "read," and "speak." And, you started Baidu Brain to gradually move toward the cognitive "insight" of prediction and judgment. You come from the future.

Thanks to Xu Jing; your insistence makes the description of the artificial-intelligence industry standard clearer. Thanks to Gu Guodong, Xiong Yun, Wang Jia, Ma Lianpeng, and other students for your ultimate pursuit and unremitting efforts. Without your supervision and internal and external coordination, this book could not have been completed in such a short period.

Thanks to Li Wenting, Cai Shuo, Ren Yiqi, Qiao Hui, Li Yingchao, Wang Linlin, Ma Xiaoxin, Chen Ming, Zhang Kaiyi, and Zhang Na for the AR presentations of this book, which endow it with magical charm of sci-fi and wizardry.

Finally, I would like to show my special appreciation to the editorial team of CITIC Publishing House for your dedication, professionalism, and rigor.

Project Team of Smart Revolution
March 5, 2017